PERMAFUTUR EDITIONS

Author : Loïc Etcheberry
Cover Page and Illustrations : Loïc Etcheberry

Commercial Use Rights for Articles and Archive Photos from BNF Gallica for Permafutur and Loïc Etcheberry

Book Layout : Gwenaëlle Castel (Instagram : @paysage.comestible) and Loïc Etcheberry

Photos of Figures 20, 55, 56, 64, 65, and 66 : Copyright-free

Illustration for Figure 58: Gwenaëlle Castel

ISBN : 978-2-9593237-4-4
EAN : 9782959323744

A heartfelt **thank you** *to Léa, Claudine, Gwenaëlle, Nathalie, Alain, and Olivier for their invaluable contributions to the creation of this second book. I sincerely appreciate their dedication, passion, kindness, professionalism, and the many remarkable qualities they brought to this project..*

We would like to express our deepest gratitude to our English-speaking readers for their support. This book has been translated from French into English with limited resources. While we have made every effort to preserve the essence and accuracy of the original text, some imperfections may remain. We kindly ask for your understanding and hope that the ideas and message shared here will inspire you just as strongly as in their original form.

———

Figure 1 : Illustration from 1984 depicting a field equipped with a large electroculture antenna.

Loïc Etcheberry is a naturally curious and passionate individual with a lifelong love of gardening. He holds a degree in environmental landscape management from the École Du Breuil, as well as training from the National Museum of Natural History and the University of Orsay. His career has taken him from the urban ecology agency of the City of Paris to the Center for Urban Agriculture Research in Montreal. From 2015 to 2022, he also taught at the Adriana Horticultural High School in Tarbes for the Ministry of Agriculture.

In 2012, he discovered electroculture and immediately recognized its immense potential for a sustainable and ecological future. A strong advocate for agroecological transition, Loïc launched the YouTube channel Permafutur in 2014 to promote these innovative solutions through lectures, garden visits, workshops, and videos. Dedicated to sharing knowledge and skills, he channels his energy into fostering hope and empowerment for a better future.

Shop, electroculture products and services :

www.permafutur.com

TABLE OF CONTENTS

PREFACE TO THE BOOK

In ancient times, the Chinese practiced what could be described as Earth acupuncture, inserting metal rods into the ground to harmonize the energy of a place with its surrounding environment. Electroculture follows a similar philosophy: working with natural energies to support and revitalize the Earth, enhancing the vitality of gardens and crops through energetic balance.

Since the pioneering work of Abbé Bertholon, Jean Nollet, and later Justin Étienne Christofleau, many researchers have sought to develop increasingly advanced and efficient aerial and magnetic antennas. Historical records are rich with experiments and observations on this subject.

In nature, trees already perform numerous electromagnetic functions similar to those of electroculture antennas: tip effects, electro-osmosis, electron exchanges between soil and air, and paramagnetic interactions. Like trees, antennas can stimulate beneficial bioenergetic exchanges, promoting plant vigor, growth, drought resistance, and resilience to disease.

The systems presented in this book vary in complexity, but once installed, they operate naturally and autonomously. As we will explore, several models can be developed, each with its own unique characteristics.

I first discovered electroculture around 2012 through Philip Forrer and his "Grail Garden." In his vegetable garden, he installs copper masts topped with zinc points, which, according to his observations, significantly benefit plant health and vitality.

The first time I saw this, I must admit I didn't fully understand how it could work. Yet the idea stayed with me, and I remember thinking: this man is onto something.

Since then, I have conducted extensive research and experiments. Where do these techniques come from? How do they work? What are their practical applications? There is a rich history behind both "lightning rod–type aerial antennas" and "magnetic antennas." These tools can be used in vegetable gardens, fields, orchards, vineyards—and even to influence water. As your understanding deepens, the wide range of possible applications becomes increasingly clear.

At first, exploring this field requires an open mind and perseverance, as it remains unfamiliar to many. Speaking about it can sometimes lead to skepticism or misunderstanding. However, over the past decade, significant progress has been made. These practices are now gaining broader recognition, largely thanks to the sharing of knowledge and the role of the internet in spreading awareness.

In this second book, following Energy Towers and Paramagnetic Basalt, I invite you to continue this journey into the vast and liberating field of electroculture, also known as energetic agriculture. You will learn how to apply these sometimes mysterious techniques in a practical and accessible way. We will begin by exploring the origins and history of lightning rod–type aerial antennas. You will also discover archival images and press articles, some of which I have personally restored and colorized.

This book is the culmination of more than ten years of research and experimentation, and I am excited to share it with you. Together, we will trace a fascinating history that helps us better understand the functions and diverse applications of aerial antennas.

Next, we will explore the use of magnets through so-called "magnetic antennas," as well as various combinations adapted to different needs. Finally, we will compare these approaches with Louis-Claude Vincent's bioelectronics. Based on these insights, we will also examine practical tools and measurement devices that can be used to analyze the effects of these unconventional systems on plants, soil, and crops.

LET'S BEGIN !

Figure 2 : Giant sunflowers in Loïc's garden.

INTRODUCTION

Electroculture (also known as energetic agriculture) is a vast and sometimes complex field that requires a holistic perspective to fully grasp its techniques and how they function. One of the key objectives here is to implement agricultural practices that respect the Environment and Nature, while promoting the germination, growth, development, and health of plants, as well as the energetic harmony of the land. This is achieved by harnessing, utilizing, and amplifying the natural electromagnetic waves and fields that surround us forces provided by Nature to Humanity, originating from the Earth, the atmosphere, and even the cosmos.

There is another form of electroculture that doesn't rely on capturing ambient electromagnetic energies but instead generates its own energy source: this is known as active electroculture. These techniques are ancient, well-established, and their effects are clearly visible. In the 21st century, they have experienced a resurgence, to the benefit of all.

PASSIVE AND ACTIVE ELECTROCULTURE :

There are two main categories :

The first method, PASSIVE, involves capturing, channeling, and amplifying natural electromagnetic and magnetic fields and currents. It requires only receivers, or 'antennas,' made from materials such as metals, rocks, crystals, wax, sacred geometry, or magnets, to harness these inexhaustible forces provided by nature. Telluric (earth) and atmospheric energies can then be concentrated in a specific area to interact with plants and crops.

For example, a simple metal stake driven into the ground and topped with a metal brush featuring pointed ends can enhance the natural flow of electrons between the sky and the earth. In the image below, taken from Lyon Horticole dated January 15, 1892, you can see one of the earliest experiments in this field. In this case, what could be described as electro-fertilizers are used to benefit living organisms. In a vegetable garden, this technique can significantly boost the health and growth of your tomatoes!

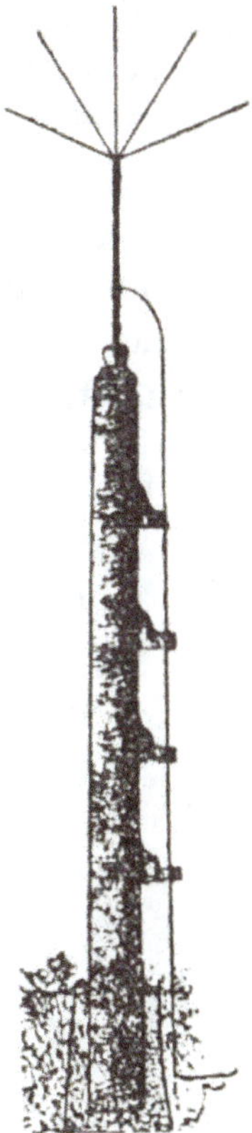

Figure 3 : Geomagnetifer, drawing by Dr. Fresnier.

ACTIVE electroculture requires an external energy source, such as a power generator, solar panels, or a direct connection to the electrical grid. This method artificially generates magnetic fields, specific frequencies, and ions that are beneficial to plants.

An example from the late **19th** or early **20th** century is that of Professor of Physics Karl Selim Lemström, who discovered a correlation between tree growth in the Arctic Circle and the northern lights. Through dendrochronology, he observed that the tree rings in the trunks were wider during years of strong solar eruptions (one ring representing one year of growth). This discovery led him to develop devices that emitted artificial currents to increase vegetable yields. Between **1890** and **1900**, he conducted experiments in Burgundy using electrified wires suspended above his vegetable garden. Under the influence of this setup, he demonstrated increased harvests of strawberries, onions, beets, carrots, and wheat. Lemström's research is detailed in his works Electricity in Agriculture and Horticulture and The Northern Lights.

These two approaches, active and passive, are different yet complementary. In this book, the focus will be on passive electroculture because, once installed, these systems operate naturally, drawing on ambient electromagnetism. There is no need for generators or external power sources beyond what nature already provides.

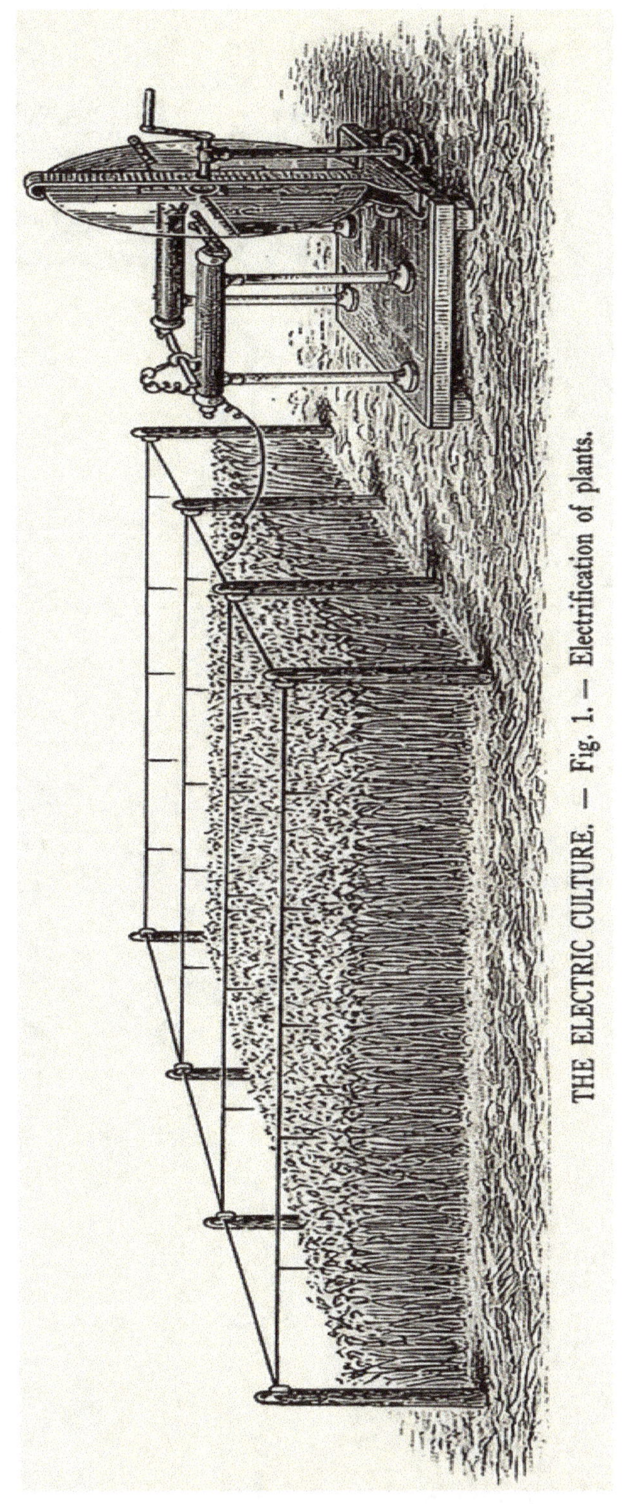

Figure 4 : Drawing of active electroculture from *Le Magasine Pittoresque*, January 1, 1894.

THIS REALLY WORKS!

or Electroculture (1) beyond all doubt

seen by MYL and PAT

with him?
CABBAGES
You should have seen the caterpillar before going through the cabbage to do electroculture !
You see, on the page opposite, there's a cow that gave for the first time, with electricity, three times more milk than she usually does!
Oh! it hurts a little— because of a very simple mosquito bite... with electroculture!
ONIONS
Oh la la! my eyes are streaming like little seas!
Not bad for keeping the size of onions!
Oh! those beautiful turnips!
These are examples of seedlings of beets treated with electroculture!
Very useful for omelettes, eggs like this, but also for laying hens!

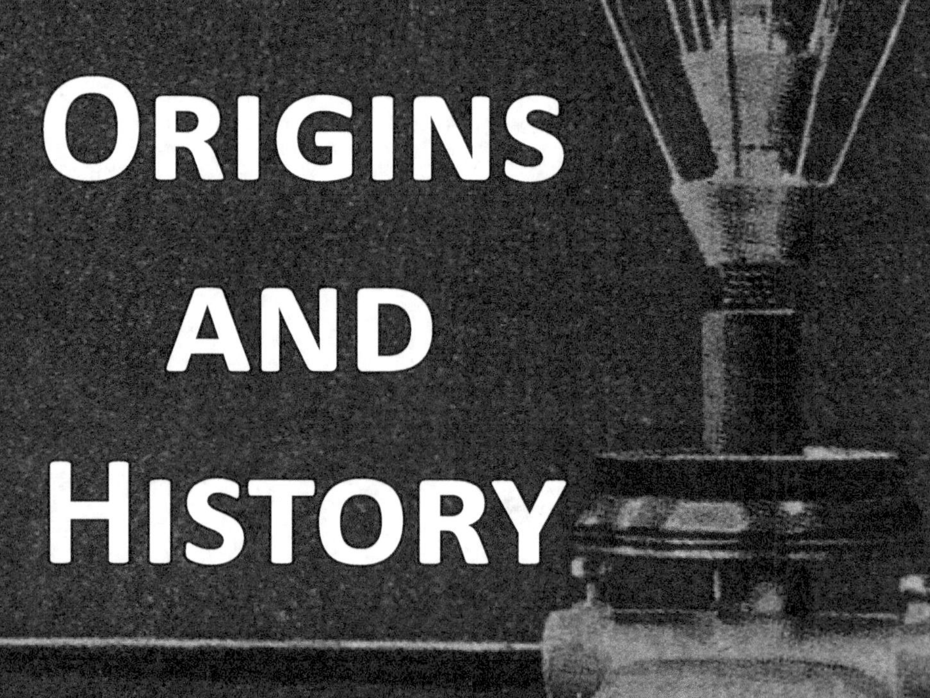

ORIGINS AND HISTORY

The evolution of modern electroculture (since the Enlightenment) has followed a progression from simple methods to increasingly complex developments and refinements over time. Early pioneers conducted experiments that laid the groundwork for future advancements.

Between **1745 and 1910**, more than 450 researchers worked on electroculture, often employing a scientific approach. This was particularly true in France, a leading nation in the field, but also in Belgium, Great Britain, Russia, Germany, Finland, and other countries. The growth of electroculture was primarily centered in Europe before expanding internationally during the **20th** century.

IN BRIEF :

At the dawn of modern electroculture, around **1790,** Brother Paulin planted a simple metal stake with pointed ends into the ground. By **1870**, the antenna was connected to an underground metal network to enhance its connection with the earth and the potential for electron exchange. In **1919**, Justin Étienne Christofleau connected the antenna to a wire aligned with the Earth's magnetic field.

The system was later modernized with the addition of a magnet. In **1930**, a new system emerged, allowing for multiple buried wires in the soil, always oriented according to the magnetic field. By **1938**, Christofleau introduced a large cast iron piece, shaped like a U-shaped magnet, containing copper and zinc elements. This was placed in the ground and connected to a device aligned with geographic north. In 1940, Couillaud observed that a simple magnet connected to a wire oriented north-south could capture magnetic energy.

Figure 5 : Drawing of the Christofleau Fertilizer from the magazine *Siroco, Le vrai journal des jeunes de France,* 1943.

IN DETAIL :

As early as the   century, with the domestication of electricity, Abbé Jean Nollet played a pivotal role in this emerging field. A member of the Academy of Sciences and a professor of Experimental Physics, he conducted notable experiments in the realm of electricity and plant growth.

According to reports, Nollet observed that plants near lightning rods appeared to grow faster. Intrigued, he carried out further experiments on the artificial electrification of seeds, concluding in 1749 that electrified plants grew more vigorously (Nexus magazine No. 69, July-August 2010).

The oldest documented work on the subject comes from Abbé Bertholon in 1783. His book discusses *The Effects of Atmospheric Electricity on Plants, Its Impact on Vegetation, Medico and Nutritivo-Electric Virtues, and Practical Applications in Agriculture*. During this period, the first 'electro-vegetometer' was developed, a simple metal antenna mounted on a mast, designed to fertilize crops by capturing electro-atmospheric charges.

In 1882, Beckensteiner mentioned the use of a geomagnetifer in the Gazette Agricole et Viticole de Lyon, in an article titled The Striking of Phylloxera. He employed an ingenious electroculture system to combat the sap-sucking insect phylloxera, which originated in the United States and was causing extensive damage to vineyards at the time.

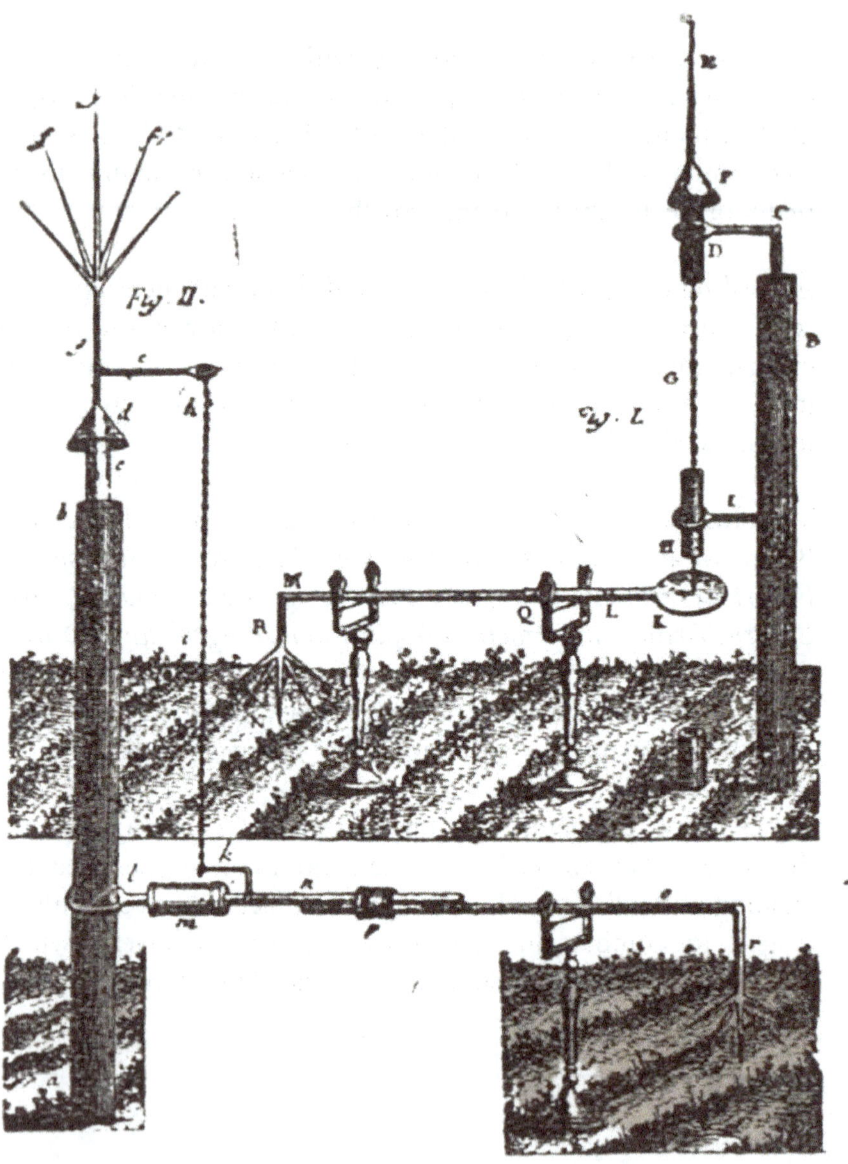

Figure 6 : Electroculture Devices by Abbé Bertholon.

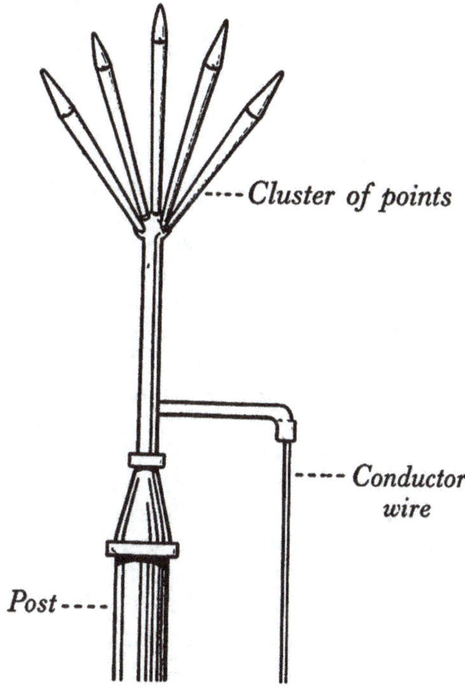

Figure 7 : Beckensteiner's Geomagnetifer from
Rustica magazine, August 17, 1930.

Later, a Russian scientist named Spechnieff perfected the electro-vegetometer and reported significant increases in crop yields: 62% for oats, 56% for wheat, and 34% for flax, according to Justin Étienne Christofleau's book Electroculture. On **May 11, 1890**, L'Ami du Peuple reported a crop surplus close to 400%! (**See article in the following pages**).

Additionally, the chronicle Le Carillon Valentinois wrote in March **1891**: «A Russian scientist has just discovered a new application of electricity that is expected to bring immeasurable benefits to humanity. This involves using the mysterious fluid to accelerate plant growth. Through this method, the scientist has reportedly achieved harvests of abundance unknown to our farmers.»

ELECTRO-CULTURE
Application of electricity to fertilization.

The application of electricity to the fertilization
of the soil, or at least the intervention of an electric
fluid developed in the soil to promote the fixation
of nitrogen in the air and its nitrification, is a cer-
rtain scientific discovery which could play a very
important role in agriculture.

Electro-culture is not a practice based on poorly
controlled experiments, but it is no less undenia-
ble than the culture by chemical fertilizers.

About forty years ago, if our fathers had been
told that the farmyard manure heap could be replaced by chemical products such as sodium nitrate,
ammonium sulfate, superphosphates, they would
have greeted these scientific novelties with the same
incredulity that many of our readers will unduu-
betly feel when considering the future role of electricity in the fertilization of soils.

But what, however, does the future have in store
for us? The electric fluid exists all around us,
nitrogen represents 4/5 of the volume of the air we
breathe ; what if one day, by means we are not
yet familiar with, it could be put to serve us in order
to capture the other and render it assimilable?
Or, as the atmosphere is to say, the entire mass
of air surrounding the earth, at an average height
of 7 to 9 myriameters, that is to say from 16 to 20
geographic locations, we see a vast reservoir of
nitrogen placed at our disposal, if we managed to fix
this nitrogen and make it assimilable.

But we must not anticipate too much and shall
come to some electro-culture experiments which
we can cite :

At the last Horticultural Congress in Paris, Mr.
Fischer, director of the Botanical Garden in War-
saw, reported, among other things, experiments con-
nducted in Russia by Mr. Spéchnief :

« Regarding the question of electro-culture, I have
considered two modes of this culture : one is the
electrification of the soil and the influence on plants
of dynamic electricity ; the other is the accumulation
of electricity in the atmosphere, that is to say the
influence of static electricity. In one case and the
other, the influence of electricity on the develop-
bnen of plants is very marked and very favorable.

To demonstrate this, I shall report two series of experiments carried out in Russia, by Mr. Spéc-chnief, and still little known outside our country. One of these two series concerns the electrification of the soil; the other, the accumulation more intense of atmospheric electricity.

To electrify the soil, Mr. Spéchnief used (in the Botanical Garden of Kiew) metallic plates $0^m.66$ long by $0^m.40$ wide. One of the plates of each pair was zinc, the other copper; the two were conne-cted at their tops by a metallic wire. These plates were placed at a distance from one another of 40 meters and almost entirely buried in the soil. Be-tween the plates connected in pairs, there was a very favorable electric current for the develo-ment of plants, since it contributed to a more complete dissolution of the substances of the soil, as well as to a more easy and more abundant absorption of these substances by plants. In short, the crops of vegetable plants, such as potatoes, carrots, beetroots, etc., were up to four times larger than those obtained under ordinary condi-tions. For other vegetable plants, electro-culture gave a surplus of 50%.

» The other series of experiments was carried out in much larger proportions and in a less favorable location with regard to climate, that is to say in the government of Pskow. In order to accumulate the electricity above the plants, the same experi-menter placed, at a certain distance from one another, metallic supports in the form of bars, converted, to insulate, into sheets of paper in scale. The supports, in the number of sixty per hectare, were united by a metallic wire and carried at their tops a collector in the form of a crown with copper teeth.

Bertholon's antenna was later optimized by Brother Paulin, the director of the Agricultural Institute of Beauvais. It is described in his book On the Influence of Electricity on Vegetation, published in **1892**. The device was named a geomagnetifer. This instrument enhances the natural electrical exchanges between air and soil, and vice versa, which are beneficial for the healthy development of plants. Vines treated with this system were found to have higher sugar and alcohol content, with ripening both accelerated and more consistent.

> **Potatoes contained more starch and were larger: «The yield per hectare was 28,000 kilos for the treated section, compared to 18,700 kilos for the untreated section. This result, achieved without special fertilization and using a common purple potato variety not known for high yields, matches the harvests of intensive farming that uses the best chemical fertilizers.»**

The antenna is composed of several elements: a wooden pole, a porcelain insulator (to isolate the rod from the pole), a rod at the top of the pole, a broom made of copper strands at the top of the rod, a conductive wire running down along the pole, special insulators for the descending wire, and a network of metal wires buried in the ground. The pole stands between 39.37 and 65.62 feet in height. The rod holding the copper broom is made of galvanized iron, measuring 27.56 inches in length and 0.433 inches in diameter, with five red copper strands, each 19.69 inches long and 0.157 inches in diameter. The conductive wire running down the pole is made of galvanized iron, 0.157 inches in diameter. Insulators are placed every 6.56 feet to ensure that electric tension is not lost due to wood decomposition. The wires buried in the ground are galvanized and connected to the main conductor. Their burial depth varies depending on the type of crops, as they must be placed within the root zone.

Figure 8 : The geomagnetifer in a field.

Issue 620 of Génie Civil, dated **April 28, 1894**, headlines: «The complete details of the successes achieved by the geomagnetifer would take up too much space to be reported.» Numerous testimonials of impressive harvests and electrocultivated crops, achieved without the use of chemical fertilizers, are also

Electro-culture by the Geomagnetifier.

The influence of electricity on vegetation has been much discussed and rather denied, following the numerous unsuccessful experiments made from time to time. Let us recall the failures of Messrs. the baron Thénard at La Ferté-sur-Crosne (wheat receiving strong dynamical currents); Naudin, at Antibes (but in a pot connected to a lightning-rod); Garolla, in Eure-et-Loir (peas electrified by Leclanché cells); Talavignes, at the school of Oudes (cabbage and chicory), etc.

It must be said, in truth, that the favorable experiments of Messrs. Petchner and Spechnew were not carried out in conditions of scientific rigor sufficient to raise conviction, especially after so many failures; also the National Society of Agriculture, in its session of January 1892, concluded, through its members Messrs. Prillieux, Dehérain, de Vilmorin, Mascart, that there was no fact certainly demonstrating the favorable action of electricity on vegetation.

Today, the proof seems to us to be made by scientifically conducted and rigorously controlled experiments. A new method, due to Mr. Paulin, has succeeded, and earned for him a medal of vermilion which the Society of Agriculture of Montbrison awarded.

The Commission, delegated by this Society, observed in the first experiment, conducted at Merlieu, in an apple orchard " which a geomagnetifier, 8 m. 50 high, had the effect of exerting its influence on a surface of twenty metres in radius. In this part of the land, the apple trees, one year of age, of extraordinary volume and vegetation, had kept until that day (September 23, 1891) a green color contrasting markedly with the neighboring parts; these trees reached up to 1 m. 47 in height and 2 centimetres in diameter.

After this first observation of the exterior vegetation, the members of the Commission laid out on this field influenced two quadrilaterals of 16 metres each of surface; then, in the rest of the land, two squares of the same area. These four plots were chosen in a place where vegetation was stronger, but representing on average either the part influenced by the field, or the other part. The 32 square metres of the influenced portion yielded 90 kilogrammes of tubers; the 32 square metres of the non-influenced portion yielded 61 kilogrammes.

» The production per hectare would therefore be 28,000 kilogrammes for the influenced portion, instead of 18,700. This product was obtained without special manure, with a variety of low-yield (ordinary violet potato), equal to the intensive crops with heavy expenditure of chemical fertilizers. The difference was even greater on October 11 : " 60 un-influenced plants yielded 38 kilogrammes of potatoes; 60 influenced plants, 63 kilogrammes. We add that the non-influenced tubers have walls, while the influenced ones have them not, and their green stems and leaves prove that the tuber was barely formed."

The analysis of the tubers of the influenced part gave less water (76.20 % against 78.60), more ash (5.30 against 5.01), less nitrogen (1.06 against 1.082), more starch (17.80 against 15.34). The potatoes of the influenced part were perfectly preserved in silos, while those not electrified presented about a fifth of their fruits rotted after a storage of about 5 months.

The second experiment took place in a vineyard, at Quérézieux, near Montbrison : « The influence of the apparatus, placed alone in August, was revealed, in a circle delimited exactly by the iron rod planted in the ground : the ripeness of the roots was more advanced and the sugars were more regular. The juice of the grapes, chosen very ripe, gave to the hydrometer and the alcoometer the following results : most influenced, sugar 16.92%, alcohol 10.4; not influenced, sugar 14%, alcohol 9.1." At the agricultural competition held at Annecy, which takes place every year, Mr. Paulin exhibited the monstrous spinach produced by his system.

The geomagnetifier, by means of which these successes were obtained, was invented by Mr. Beckeristener and greatly modified, simplified and perfected by Mr. Paulin.

The actual geomagnetifier (Fig. 1) consists of a resinous perch, about 12 to 18 metres high, hollowed out and painted with oil or better still, with several coats ground into the earth: a rod of iron in porcelain T, analogous to the insulators of telegraphic poles, supports at its top a galvanized iron rod F, $0^m.70$ long, ended by a balai (brush) of five branches S whose copper tips are $0^m.35$ long. From the iron rod F descends a galvanized wire G, held against the perch by special insulators I I, and intended to carry the electric current by the wires BC and ED in the portion to be influenced (rectangle or circle). The wires are buried at a variable depth according to the crop (0.15 to 0.50), so as not to injure the roots.

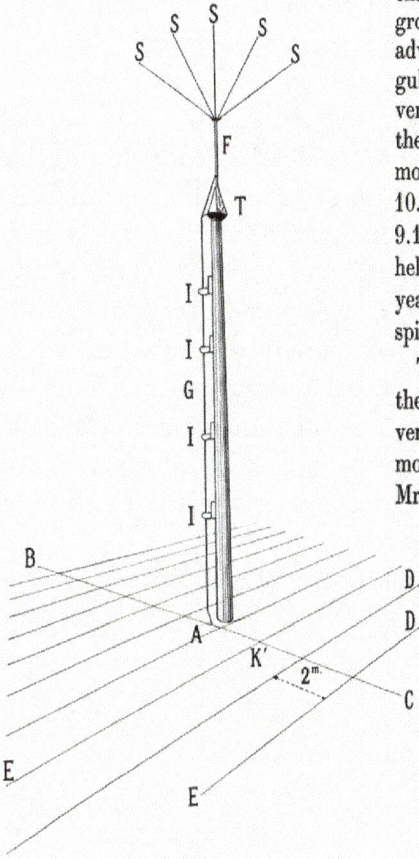

Fig. 1. — Geomagnetifier.

Mr. Paulin observed that the influence is felt exactly up to 1 metre from the wire; he therefore advises the arrangement shown in the figure herewith (AC, AB = 25 m. each; EK, KD = 25 m. each; KK' = 2 m.); hence, 4 geomagnetifiers per hectare must be used.

It is to be desired that the geomagnetifier be experimented with in different soil conditions, crops, climates, etc., so that the question of electro-culture may be perfectly elucidated : perhaps it will then be possible to give a satisfactory theory of the facts observed, but at present they remain inexplicable.

C. CRÉPEAUX.

Figure 8 bis : Excerpts from Issue No. 620 of *Génie Civil,* April 28, 1894.

In **1911**, the Bulletin of the Royal Botanical Society of Belgium reported on the experiments conducted by Lieutenant Basty of the 135th Infantry Regiment of Angers (pages 27 to 31).

At the time, his techniques promised a bright future for vegetable gardens and small-scale crops. More details can be found in the book On the Electrical Fertilization of Plants by Fernand Basty, dated **June 27, 1910**. He conducted experiments over seven years before seeing his efforts crowned with success. Control plots, alongside those equipped with electroculture systems, were used, and around thirty species exhibited beneficial and striking effects. A photograph of his experimental garden in Angers is included in the book (on the following page). He named this experimental site the *Bertholon Garden*, likely in tribute to Abbé Bertholon, who had previously worked in this field.

Here are two examples of the lieutenant's techniques:

1/ **Electrification of seeds before sowing, using a direct current of 0.4 amperes and 6 volts.**

2/ **The use of aerial-type antennas, such as electro-captors, dynamo-captors, and small lightning rods. Their lengths varied depending on the size of the plants being treated (2 meters for cereals and 0.80 meters for strawberries). These antennas were designed to enhance natural electrical exchanges between the soil and the air, directly interacting with the cultivated plants. Early ripening, abundance, and improved quality were observed: «Spinach, peas, strawberries, and other produce were harvested by May 15, while three weeks later, the control plants had yet to yield anything.»**

Figure 9 (next page) : Overview of the 'Bertholon' garden.

General view of the " Bertholon " Garden on June 27, 1910

Cliché Lechattier, Angers.

One can remark: *a)* The devices: 1. Electro-captor; 2. Dynamo-captor; 3. Small para-tonnerres.
b) Influence of the devices on: Cabbage (4), see its control (5); Beans (6), see its control (7); Barley (8), see its control (9).

§ 3. *Description of the Electro-Captor F. B.*

In 1907, we used for our experiments a geo, which we called the *Electro-Captor*. This very simple device, within the reach of all purses, could be easily built by anyone, even the least skilled person (provided, however, that he knows how to solder); it gave us excellent results.

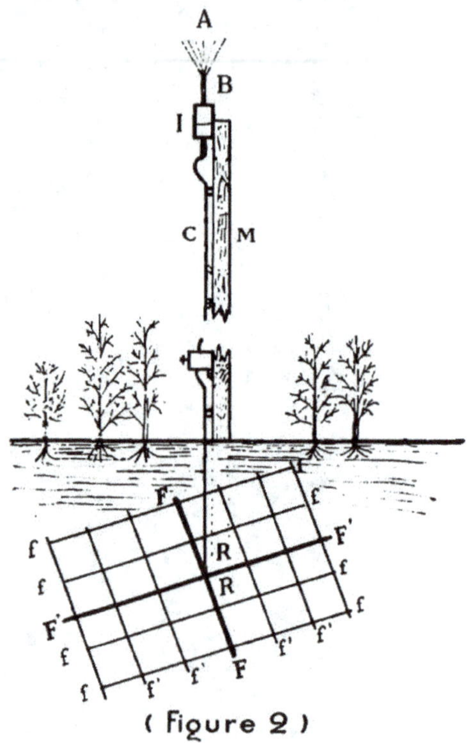

(Figure 2)

Here is its description in a few lines :

At the top of a mast (M) (as high as possible), firmly planted in the ground, we place an insulator of porcelain or glass of a special model.

This insulator (I) has the shape of a hollow cylinder following its length. Inside the insulator will be inserted the base (B) of a needle (A) made of copper wire, the lower wires of which will

Figure 10 : Excerpt from page 37 of On the Electrical Fertilization of Plants, June 27, 1910, by Fernand Basty.

Thanks to the growing accumulation of positive results, enthusiasm for electroculture spread rapidly across Europe. As a pioneer in the field, France played a central role in its development and acceptance within scientific circles.

From October **24 to 26, 1912,** the first International Congress on The Influence of Electricity on Plants and Possible Applications in the Agri-food Industry was held in Reims. Among the attendees were François Kövessi, Professor of Botany and Plant Physiology at the Higher School of Hungary (who studied the effects of electricity on plant germination), Henri Bresson, Laureate of the Academy of Sciences, and Colonel Pilsoudski, among others. The latter conducted experiments at the Saint Petersburg Botanical Garden, where he made the following discovery by chance:

«While performing certain tasks, we had to immerse two electrodes in a canal used for irrigating nearby fields. The electrodes were connected to a powerful voltaic battery of about 200 elements: one electrode was zinc, and the other copper. After the work was done, the electrodes were left in the canal. For reasons unknown to me, the conductive wires were eventually connected. On the banks of the canal, poplar trees were growing, which typically take 10 years to become strong enough for construction work. The forgotten electrodes generated an electric current, and when we returned some time later, I observed that the poplars planted between the two electrodes were twice the height of the neighboring poplars.»

The benefits for plants were clearly visible at the time, and many were demonstrated during the congress, including the use of electric treatments for the sterilization of liquids and food products, as well as for the rectification and aging of alcohols.

By this time, the range of both active and passive electroculture techniques was already quite extensive.

1st INTERNATIONAL CONGRESS

of Electro-Culture

AND

APPLICATIONS of ELECTRICITY

to Agricalture, Viticulture, Horticulture and Agricultural Industries

HELD AT **REIMS**, FROM OCTOBER 24 TO 26, 1912

in the presence of the Official Delegates of the Ministry of Agriculture,
the Academy of Sciences, and the Governments of Belgium,
Hungary, Luxembourg, Mexico and Russia.

PROCEEDINGS

BY

A. Ph. SILBERNAGEL

Consulting Engineer
Director of " *Le Génie Rural* " and " *L'Électroculture* "
President of the Congress Organizing Committee

with the collaboration of

Fernand BASTY

Editor-in-Chief of " *L'Électroculture* "
Laureate of the National Society
of Agriculture of France
General Secretary of the Congress

Jean ESCARD

Civil Engineer
Laureate of the Society of Encouragement
for National Industry
General Reporter of the Congress

PUBLISHED BY ÉDITIONS DE TECHNIQUE AGRICOLE MODERNE
OF THE CENTRAL OFFICE OF " LE GÉNIE RURAL
OF MOTORCULTURE AND ELECTRO-CULTURE
58, Boulevard Voltaire, 58
PARIS

Fascicle No. 3

The Encyclopedia of New Inventions in Science and Work, published in **1926**, dedicated 30 pages to the topic of electroculture. It discussed the use of atmospheric electricity on plants, referencing the experiments of Duhamel, Monceau, Nollet, Bertholon, Grandeau, and Leclerc. The encyclopedia also explored the use of voltaic current, light radiation, discharge by effluvia, and the electrification of seeds before sowing. Observations of positive, neutral, and sometimes negative effects on the plant world were thoroughly documented. Mistakes made during experiments provided valuable insights and helped clarify certain challenges related to implementation.

Figure 11 : Active Electroculture Experiments in Paris in 1921.

Following some partial failures in earlier experiments, cultivation conditions were rigorously controlled in **1921** in Bellevue (Paris). The plants were grown in plots exposed to an electric field generated by electrifying a metal net placed above the crops.

Electrocultivated and control plots were set up under similar conditions, with the only difference being the presence of electricity. High-voltage current was conducted through a network with both aerial and underground components. A metal plate measuring 10.76 square feet was buried 3.28 feet deep, in contact with the soil. Moisture near the plate was maintained, and the current's characteristics were monitored using an ammeter and a voltmeter. The electrical effluvia were applied daily for four hours, avoiding periods of strong sunlight, which had been identified as unfavorable in previous experiments. Lettuce, chicory, cabbage, carrots, potatoes, beets, celery, leeks, onions, and tobacco were compared with and without treatment. The results clearly demonstrated the favorable influence of active electroculture via effluvia. Throughout the experiment, the visual differences between treated and untreated plants were evident.

Some comparisons between electrified crops and control crops :

4 season lettuces : 15.43 lbs / 8.44 lbs
Milan cabbages : 28.66 lbs / 21.93 lbs
Carrots : 22.05 lbs / 15.43 lbs
Potatoes : 66.14 lbs / 48.5 lbs
Leeks : 12.35 lbs / 6.39 lbs

Frenchman Justin Christofleau, Knight of the Order of Agricultural Merit and member of the Society of Scientists and Inventors of France, was one of the great pioneers of the **20th century.** A talented engineer, he developed numerous electroculture systems that were exported internationally. Unfortunately, he fell into obscurity after World War II.

Christofleau created many effective devices for electroculture and was even the inventor of the electric bread-kneading machine! Between **1905 and 1939**, he filed an average of 1 to 5 patents per year. His writings were translated into several languages, and one of his books (in English and Spanish) was stamped by a bookstore in Canberra, Australia. He was also photographed in China with local farmers.

In addition, Justin Christofleau significantly advanced electroculture systems, developing practical and concrete applications for agriculture. His patents can be found at the Ministry of Commerce and Industry, including patent number 829,789 for the Christofleau electromagnetic fertilizer. This device was reportedly distributed in over a million units worldwide. In his book Increased Harvests and the Saving of Diseased Trees through Electroculture, he demonstrates how it is possible to farm without chemical fertilizers, using what he called 'terro-celestial electromagnetic fertilization'.

The Paris magazine *La Liberté* from Saturday, November 3, 1934, features Justin Étienne Christofleau with the headline: *Un tour au jardin de Gulliver* In the article on the following pages, it recounts some of his achievements.

Figure 12 : Justin Étienne Christofleau with rye reaching 6.6 feet in height.

Queue-lez-Yvelines. A single street in the shape of a crescent quarter-moon. A rooster crows in the street. At the edge of a window, chrysanthemums bathe in a milk-pot from which it rains. Paris is half an hour from the end of the world.

At the far end of the village, a property with white shutters, sheltered by large trees. This is where Mr. Justin Christofleau, a sorcerer, scholar, thinker, works. He welcomes you in a large dug-out room painted red, where the clock ticks without being wound. And yet!

Near the door, a three-metre-high plant dominates you. Some specimen of tropical flora? Above all, it is a maize from the Landes, from the last harvest. Its ears are there, in a box: each is ninety-nine centimetres long and a half, and weighs two hundred and forty grammes. The ears of Caragua maize or cheval-tooth measure twenty-six centimetres and weigh six hundred grammes. Yet, from memory, I never knew whether he had died in Queue-lez-Yvelines.

O Gargantua, here is a potato. It is twenty-six centimetres long, twenty-five in circumference. Rough carrots: ninety-nine centimetres without the tops, and a total length of one metre. A circumference, finally, of twenty-eight centimetres.

Beetroots, insolent in their dye, are ninety-nine centimetres long, fifty-three in circumference.

— I have planted only ten tomato plants, said Mr. Christofleau. They have given three hundred kilos, I could have maintained the whole country... We eat them from the beginning of August, and there are still some in the vegetable garden.

"This summer, at least two thousand kilos of peaches were picked, and yet we gave a quantity to the children."

*
**

"But let us go and throw a glance at the garden..."

It seems that the fine season is sheltered on this hectare. Perfumed roses, nasturtiums, raspberries, strawberries loaded with red fruits. In the month of November, not a sign of frost.

with the leaves of a very dark green, almost black thick like cardboard.

On a very old apple tree, the shoots of the year reach two meters.

A snail is as beautiful as you would find in Bourgogne.

The gardener is establishing a new apparatus of electro-culture, without poles, masts or antennas.

— It can intensify a strip of land thirty-six meters wide by one thousand and even two thousand meters long if necessary. One can measure the electricity that one draws from the ground as if it came from the power plant sector. In fact, the apparatus is connected to a milliamperemeter, and we note a deviation of two hundred milliamperes.

In summer, the electric tension is even stronger.

In Morocco, it reaches four hundred and fifty.

*
* *

Here, the monstrous vegetable is the king. In the square room, the clock strikes all the hours on the wall that hides the shadows of evening. Here is a cabbage in which is surely born the child who holds it. Here is a turnip in wooden shoes and sabot, and a carrot that crosses the hands of a peasant woman. Three-book pears. Spinach leaves, large endives like parasols. And here — greetings! — the diploma of Agricultural Merit, in the place of honor among so many photographic ex-votes, recognition of vegetable growing to the Unknown Fairy.

— Electricity, that is the source of life.

I am sixty-ten years old. I work sixteen hours a day and I sleep only two hours each night, but... in the middle of a magnetic field.

"Physics, chemistry, mechanics, alone, I have learned all and a locomotive no longer embarrasses me to build more than a watch.

From morning till night, mechanic by day, mechanician by night, it is I who must be later the inventor and the propagator of mechanical petrissage, for which the National Society of Encouragement for National Industry awarded me, in 1908, the gold medal...

" For there—fifty-nine years already—I leave my dindons and my Touraine, the whole of my fortune: three francs in a pocket!"...

Juliette LARTIGUE.

The system below is from the magazine Rustica, dated August 17, **1930**, and comes from the Bibliothèque Nationale Française Gallica. It illustrates one of the sophisticated devices invented by Christofleau, which was generally very effective. This system could be connected to fruit trees, vegetable crops, vineyard trellises, and more.

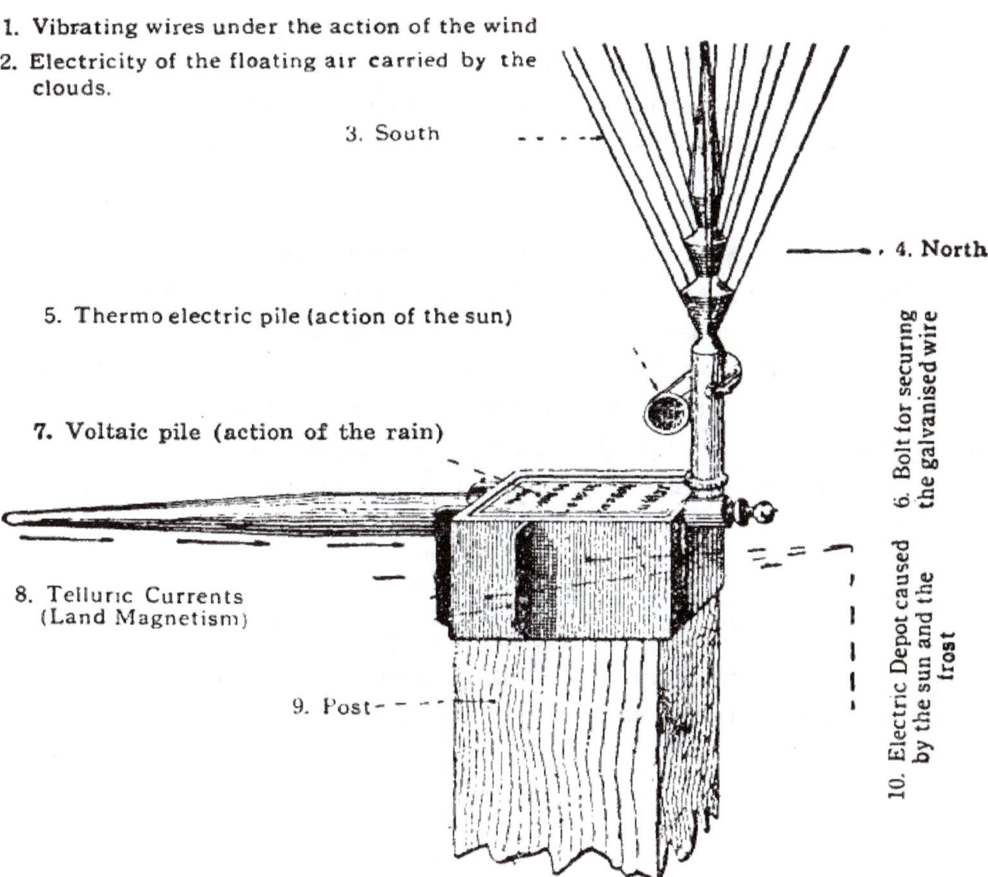

Figure 13 : Electroculture device in Electroculture, by Justin Christofleau.

Electricity and magnetism were harnessed using this ingenious system. Atmospheric electricity was captured by a needle positioned perpendicular to the ground, with multiple metal rods erected along its length. The horizontal needle captured terrestrial magnetism and was aligned with geographic south, following the Earth's magnetic field. It was made of ferromagnetic materials. Solar action was also considered, with fins arranged on the sides to create temperature differences and induce an electrical deposit. The natural battery also operated using a copper-zinc combination. The effects of frost, rain, and wind were also taken into account. For example, wind caused the antennas to vibrate, capturing electricity from the air. Notches on the voltaic pile allowed electrons to be extracted from water droplets.

This system needed to be at least 21.3 feet high and was grounded by various means. On the right side of the central box was a bolt (see previous page) used to connect a galvanized iron wire. This wire was then attached to a vineyard trellis, the base of a tree, or a buried wire in the soil. To facilitate energy exchanges, critical details had to be considered, such as ensuring the galvanized wire along the mast was insulated to prevent energy loss. Insulation was achieved using ceramic pieces.

Thanks to his fertilizer, Christofleau produced extraordinary crops without chemical fertilizers, simply by harnessing the natural electromagnetic forces provided by nature.

«*If fertilizers are used to boost growth, one should not assume that chemicals directly influence vegetation. The facts are as follows: all decomposing chemical substances emit an electric current; it is this electric current, resulting from the decomposition of fertilizers in the soil, that provides the vegetation with the necessary energy for intense growth. The elements from the atmosphere nourish plants far more than the soil itself, reinforcing our assertion: if chemical fertilizers increase production, it is because their decomposition in the soil generates an electric current that amplifies the one naturally present in the atmosphere.*»

J. Christofleau continues: «*The capture of atmospheric electricity for the benefit of crops is therefore an invention of the utmost importance... Only skeptics would refuse to use this natural force that costs nothing. To intensify yields, one should not wait for nature's forces to naturally fertilize the soil: they must be captured, channeled, and directed to where they are needed.*»

This is exactly what Justin Étienne Christofleau's devices are designed to achieve. The results are equally impressive: in a field that was neither irrigated nor fertilized, hay grows to 7 feet 1 inch in height, clover reaches 5 feet 3 inches, and potato plants have stems 6 feet 3 inches tall, with each plant producing 30 to 35 tubers weighing between 1.1 and 2.2 pounds, all of exceptional quality. Carrots grow to 18.9 inches, beets reach 18.1 inches in length with a circumference of nearly 16.9 inches, pears weigh 2.6 pounds, and vines affected by phylloxera were healed and rejuvenated.

A revolution in agriculture was the headline of one newspaper at the time, and indeed it was...

*He achieved astonishing results :
«…leeks with a circumference of 11.8 inches, carrots weighing 4.4 pounds and 19.7 inches long, endives measuring 33.5 inches, cabbages with a diameter of 13.8 feet, pears weighing 2.6 pounds each, and parsley growing to 23.6 inches in height…» recounts Alexis Danan in Paris-Soir of January 11, 1932.*

However, caution is advised. After spending hours poring over the archives of the BNF (French National Library) Gallica, one occasionally comes across old articles that present counter-examples. In a few instances, though not the majority, I have found writings from that era suggesting that electroculture experienced some failures. Indeed, in certain contexts, the results were not as expected. There are countless factors to consider, and nature cannot be fully controlled. Perhaps, in some cases, the electroculture apparatus did not work as intended—or at all? There is still much to uncover...

This uncertainty is also reflected in Bulletin of Research, No. 210 from **March 1937**, published by the College of Agriculture and Mechanical Arts of Iowa. An experimental agricultural station was set up, and writings by Charles S. Dorchester, such as «The Effect of Electric Current on Certain Crop Plants,» still remain.

Figure 14 (next page) : Aerial Antennas from Electroculture Experiments in Iowa, 1930-1932

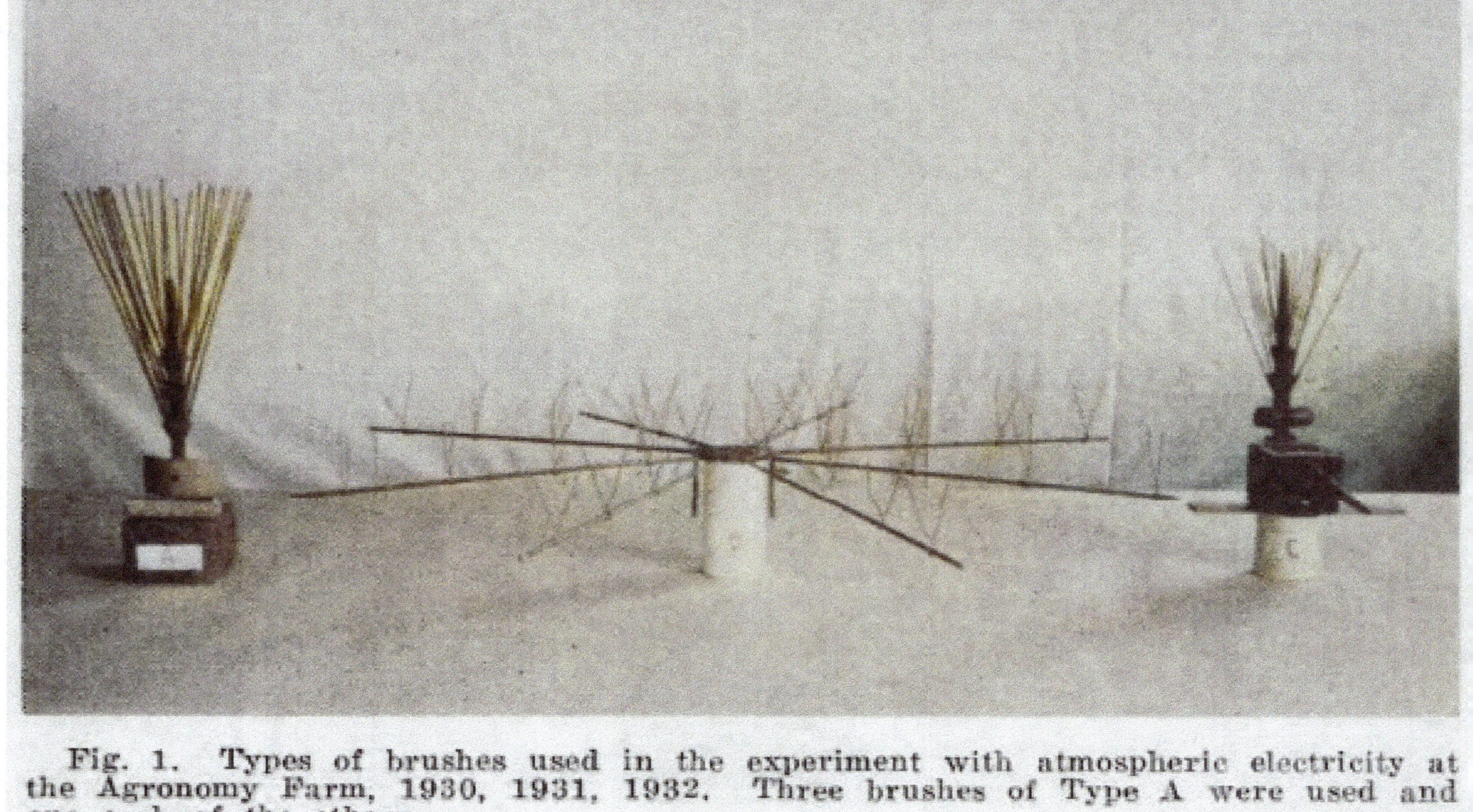

Fig. 1. Types of brushes used in the experiment with atmospheric electricity at the Agronomy Farm, 1930, 1931, 1932. Three brushes of Type A were used and one each of the others.

Type A—from C. Vincent, Omaha.
Type B—from L. W. Butler, Iowa State College.
Type C—from Justin Christofleau, La Quene les Yvelines, France.

The aim was to compare the effectiveness of different aerial antennas for electroculture over a period of three years. However, it seems there were challenges in definitively validating these techniques at the experimental station.

I invite you to read the full report of the study for a more detailed understanding of the accounts from that time.

Figure 15 : Experimental Station in the United States from 1930 to 1932

After World War II, starting in the **1960s,** the Green Revolution emerged. This was a political and agro-industrial push to increase agricultural productivity. The use of chemical inputs in farming became more widespread. The landscape changed with land consolidation, the removal of hedgerows, and the expansion of open fields, paving the way for intensive farming and the use of petrochemicals.

The primary subsistence goal of family farming has shifted to commercial export agriculture, benefiting financial markets and globalization. Farmers have become increasingly dependent on the international economy and the stock market. This period of increased productivity through chemical fertilizers also brought significant consequences: disruptions to human and environmental health, nitrate surplus, pollution of aquatic ecosystems, soil erosion, loss of soil nutrients and humus, decline in biodiversity, and a drastic reduction in the number of farmers. The list is long and well-known today.

Despite this, electroculture is no longer in the spotlight. However, it has endured over several centuries, with its share of failures and successes, but more importantly, with much hope and promise.

It occasionally resurfaces, albeit more marginally. «Lettuce with a circumference of 39 inches; leeks as thick as an arm and 47 inches long; wild strawberries as large as walnuts; dwarf peas 39 inches high with pods 8 inches long...» This gigantism may be possible thanks to a particular technique presented in Rustica magazine in **June 1975**. The article describes a lightning rod-type antenna connected to a flat spiral buried or covered with sand and earth.

This device purportedly harnessed the Earth's magnetic field along with telluric and aerial energies. To achieve this, a circular trench 20 inches deep and about 16 to 20 feet in diameter needed to be dug. In the center, the antenna stood 23 to 26 feet high, mounted on a wooden or aluminum pole. From what I have read in the documents I found, proper insulation was also necessary for the two descending wires along the pole. The telluric waves were to be concentrated on a spiral that acted as a trap for the magnetic field's lines of force.

A continuous current is then established along the spiral, creating a magnetic field whose strength depends on the diameter of the spiral and the number of turns. The entire setup is assembled, welded, insulated, and then buried in the ground, starting with a layer of sand completely covering the spiral, and finally topped with soil.

Figure 16 : Spiral Antenna in the June 1975 edition of Rustica magazine.

Following the installation of this system, the device was reported to start working immediately. According to the article, no fertilization was required, and the effective range was approximately 98 feet. Silica-rich soils were more favorable than clay soils. If the wire was buried at a depth of 19.7 inches, it was only to prevent damage during plowing. The 1975 magazine suggested that harvesting strawberries or lettuce on a stormy day was not recommended.

Indeed, each reading of past texts provides new insights into these techniques, which today enable us to achieve increasingly positive results and more effective systems. This is the only document where I encountered this type of antenna. One person even mentioned that a friend had built a similar setup about twenty years ago and reportedly achieved excellent results!

The technique of aerial antennas reappeared once again in Rustica magazine, issue 456, in 1978. Since then, there have been no further mentions of this technique in that prestigious plant publication.

On May 28, 1984, Martine Quereyl, a future Doctor of Pharmacy, defended her thesis on the effects of passive electroculture on plants. This thesis was presented at the University of Limoges, Faculty of Medicine and Pharmacy. Her research and investigations focused on analyzing the influence of passive electroculture systems on the active principles of plants.

Three species commonly used in pharmacopoeias were studied: peppermint, belladonna, and Jimsonweed. «The advantage of these plants is not only to observe the effects of electroculture on growth but also on the content of active principles,» she noted in her thesis.

The study adhered to scientific protocols and standards: «Once drying was completed, we performed a new weighing to determine the weight of the dried plants.»

Each part of the dried plant was then ground into powder, and various analyses were conducted: moisture content determination, quantitative analysis (identification reactions, chromatography), and measurement of active principles.

> *The observations show that electro-cultivated plants, compared to the controls, are not only visually larger but also richer in active principles.*
>
> ***For example :***
> *27% more essential oil in mint ;*
> *57.5% more alkaloids in datura leaves ;*
> *25% more seeds.*

However, her work was not widely publicized, and her thesis did not receive the recognition it deserved, both for her research and for electroculture, according to Nexus magazine **(August 2010)**.

After the post-war period, only a few rare studies persisted, and electroculture fell into denial and near total oblivion.

Today, we have the Internet. Although it is not perfect, as I often say, «with a hammer, you can build or destroy.» It is not the tool itself that is good or bad, but how we use it. Information is now widely accessible, and the whole world is connected. The sharing of old data is made easier, and historical documents and archives are resurfacing. Personally, it is partly thanks to this tool that I have been able to learn so much in such a short time. It is through this tool that I can share knowledge and experiences with people from around the globe. It is a real opportunity. Today, it is very possible to revive this vast field of electroculture for the benefit of all. More and more people are eager to experiment, research, and share.

We all recognize the different possibilities offered by electroculture because we share the desire to revive it. In a sense, we are like neurons in a collective superbrain working toward this goal. Together, we form the electroculture community: gardeners, researchers, farmers, vineyard owners, market gardeners, scientists, arborists...

Renewal is both underway and possible. Currently, there is a genuinely positive evolution in this field. Numerous studies and experiments clearly demonstrate that the potential and solutions are there.

Everything needs to be (re)developed, shared, refined, and improved. Our era also offers the opportunity to conduct more advanced analyses using precise measuring tools. Science is progressing as well, helping to reassure the general public and those who remain skeptical. The study of internal and external plant metabolism is now more advanced thanks to plant electrophysiology, allowing us to better interpret the phenomena at work in the plant world, soil, crops, and pests. In agroecology, many effective and well-established solutions already exist.

I am deeply convinced that it is possible to sustain humanity without the use of pesticides or chemical fertilizers. 21st-century civilization faces a major challenge: shifting paradigms and reconnecting with our planet and all forms of life. Permaculture, agroforestry, biodynamics, direct seeding under cover, agrosylvopastoralism, electroculture... Everything we need is already here, within reach, to move forward in this direction.

We are all actors in the world of TOMORROW, and the sustainable future is being written TODAY.

ELECTROMAGNETIC

ECOLOGY

Matter = energy
Electro-atmospheric circuit
Tension lines
Direction of aerial current
Tellurism

MATTER IS ENERGY

A physical object is composed of atoms and particles that are constantly shifting position. A solid appears to us as a static whole, but it is actually the result of continuous vibrational motion. Matter is in motion; it is energy. When we zoom in at the molecular level, whether inside the cells of a living being or at the core of the hardest rock, we find atoms. Atoms consist of a nucleus around which electrons orbit, and these electrons are always in motion. A simple example is the hydrogen atom, with one proton (nucleus) and one electron. Nuclei can have a single proton, as in hydrogen, or multiple protons. In the latter case, there will also be an equal number of neutrons.

Figure 17 : Schematic representation of an atom with 3 electrons (-), 3 protons (+), and 3 neutrons.
Its overall charge is neutral.

In a neutral atom, there are as many protons (+) as there are electrons (-), which ensures the overall neutrality of the atom. Atoms contain both positive and negative charges: protons (in the nucleus) are positive, electrons (in the outer region) are negative, and neutrons are neutral.

An ion is an atom whose electric charge is not neutral. Ions form through interactions with other atoms. In some cases, electrons are exchanged, and an atom can lose or gain one or more electrons. As a result, the number of proton(s) (+) in the nucleus is no longer equal to the number of electron(s) (-) surrounding the nucleus.

Figure 18 : Gaining 1 electron (-) = the overall charge becomes negative.

If there is a gain of electrons, the atom becomes an **anion**, a negatively charged ion.

For example, if there are 4 electrons (-) and 3 protons (+), the overall charge of the atom is negative (-).

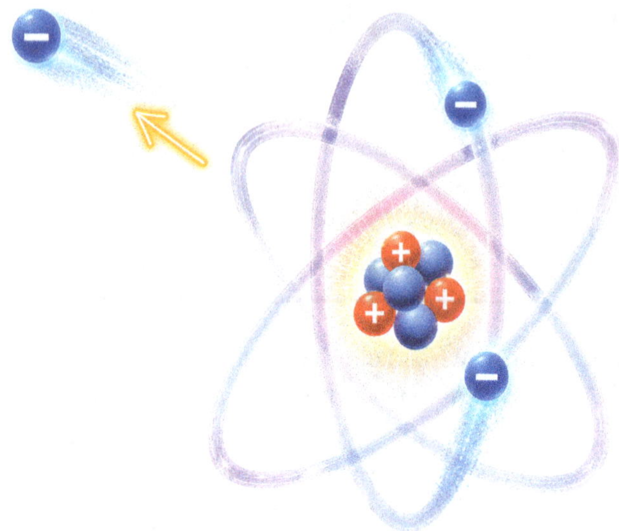

Figure 19 : Losing 1 electron (-) = the overall charge becomes positive.

> If there is a loss of electrons, it becomes a *cation*, a positively charged ion.

In the example above, 2 electrons (-) and 3 protons (+) are represented, resulting in an overall positive charge (+).

This may seem complex, but understanding these concepts helps us recognize the presence of positive (+) and negative (-) charges in living organisms and the environment. Moreover, these electron transfers are utilized in electroculture to create battery effects or potential differences between metals, for example.

As a reminder, matter is energy, and it largely consists of empty space. There is a vast amount of space between the nucleus and its electrons.

THE GLOBAL ELECTRO-ATMOSPHERIC CIRCUIT

There are two main conductors in the global terrestrial system: the Earth and its atmosphere. Together, they form a kind of giant spherical capacitor. According to the highly regarded book Atmosphere, Ocean, and Climate by Delmas, Chauzy, Vertraete, and Ferré, the Earth carries an overall negative charge of -550,000 coulombs, while the atmosphere holds a positive charge of +550,000 coulombs.

> The *coulomb* is a unit of measurement that defines electric charge. To visualize it, think of it as a quantity of available electricity. The higher the value, the more energy is present.

This does not mean that there are no positive charges in the ground or negative charges in the air... They exist, but if we total all the electric charges in the air and those in the ground, the air is overall positive, and the ground is negative. The atmosphere-Earth system tends to electrical equilibrium, as nature seeks balance.

Additionally, this arrangement of positive and negative charges creates an average difference of 300,000 volts between the ionosphere and the Earth's surface, with the maximum potential at the ionosphere (the upper layer of the atmosphere). This means that if the ground is at 0 volts, the top of the ionosphere is at 300,000 volts. This also creates a rate of change in electric potential that varies with altitude, expressed in volts per meter (V/m). In clear weather, without clouds, this electric field has an intensity of 100 to 120 volts per meter. (1 meter = 3.28 feet)

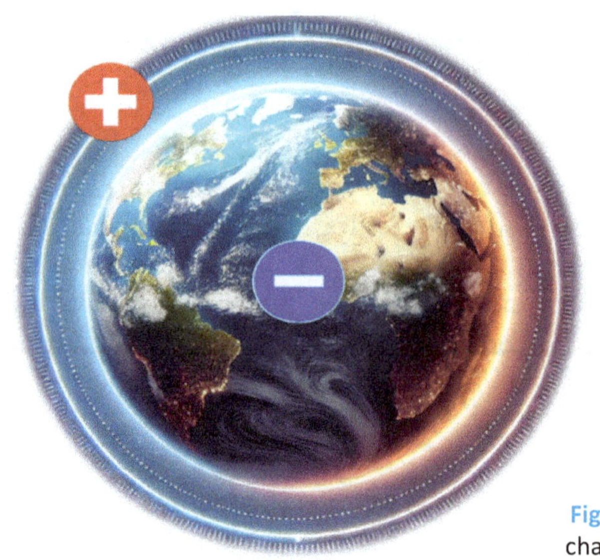

Figure 20 : The overall electric charge of the Earth is negative, while the atmosphere is positive.

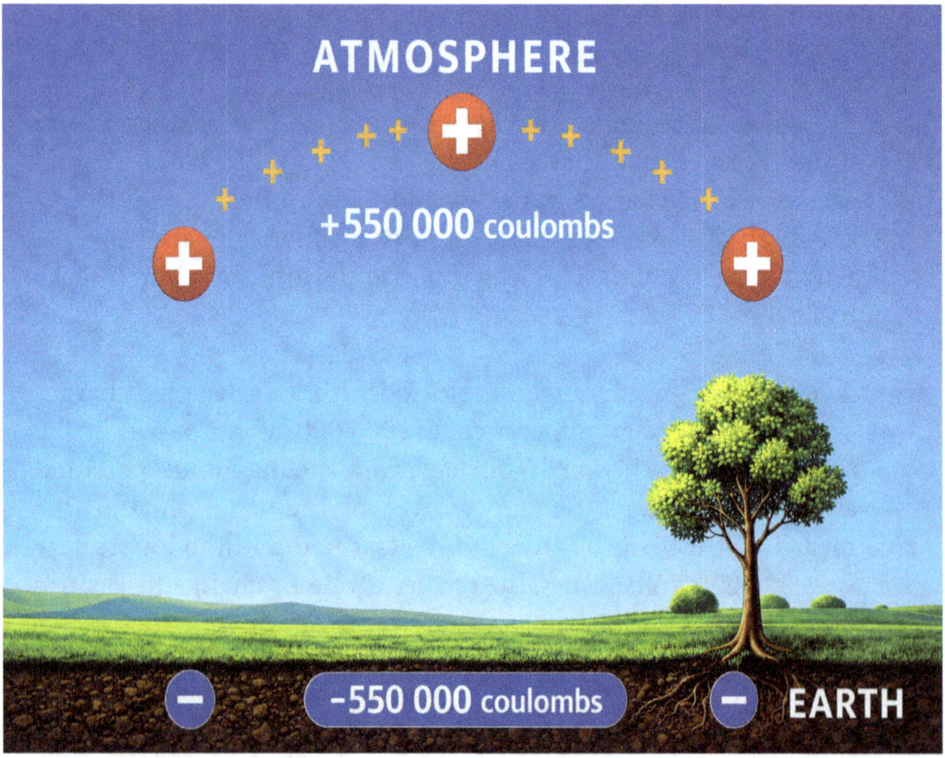

VOLTAGE (OR TENSION) LINES

This voltage-per-meter (or tension) system creates parallel electric field lines near the ground. At a height of 16.4 feet (5 meters), there are theoretically 500 to 600 volts, compared to the ground, which is at 0 volts. In reality, these lines are more or less horizontal. For instance, plants can cause the lines to bend depending on their shape.

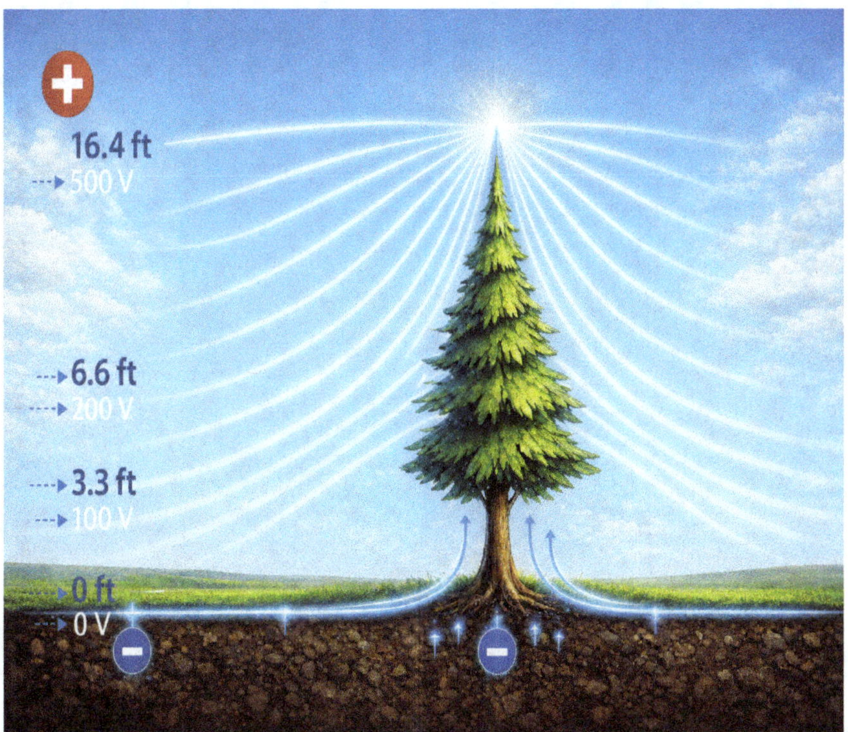

Figure 21 : The tree distorts and redistributes the atmospheric electric field lines.

Another example: a voltage of 100 volts (normally located at a height of 3.3 feet in open terrain) could be found at a height of 6.6 feet due to the presence of a plant that is 6.6 feet tall, as the plant will raise the field lines. This phenomenon can also occur with aerial electroculture antennas. Moreover, the higher the antenna is positioned, the more the field lines can theoretically curve, increasing the potential difference between the bottom and the top of the antenna.

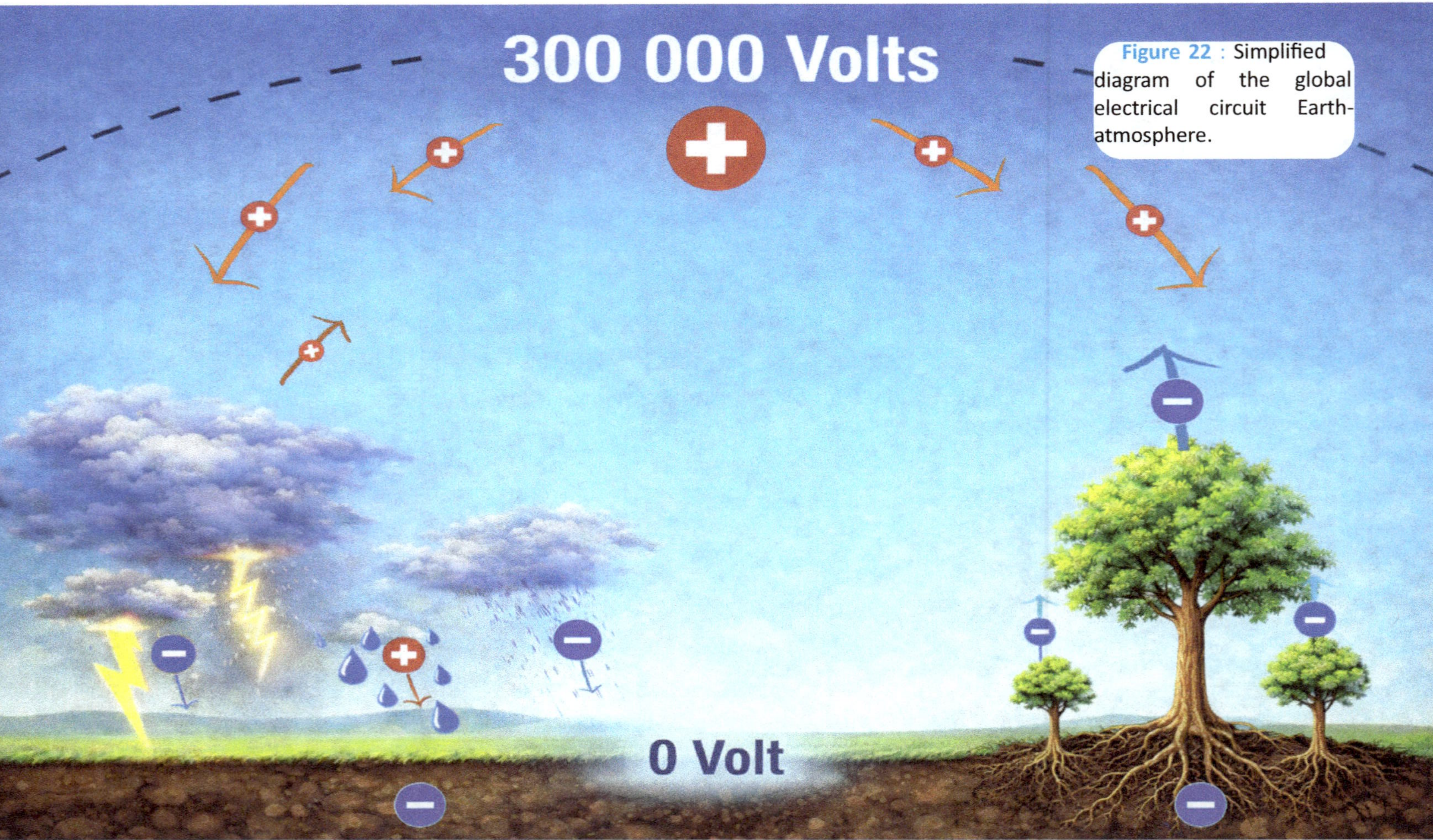

Figure 22 : Simplified diagram of the global electrical circuit Earth-atmosphere.

DIRECTION OF ATMOSPHERIC CURRENT

Moreover, there are different directions of currents between the Earth and the atmosphere, flowing both from top to bottom and from bottom to top. These currents are measured in amperes per square meter. For example, the Earth continuously discharges electrons due to the point effect, which rise through trees and are ejected from their tips or trichomes (see Energy Towers and Paramagnetic Basalt by Loïc Etcheberry). The sky, in turn, continuously releases a portion of positive electricity. The Earth receives increasing amounts of positive electricity and should, theoretically, become less and less negative. This gradual reduction in the Earth's negative charge is further reinforced by another effect: rain, which brings down both negative and positive ions, with positive ions being the most abundant and landing on the Earth.

Yet, despite these processes, the Earth's charge of -550,000 coulombs fluctuates little and remains relatively constant. This is because several mechanisms work to maintain the negative charge and balance within this giant electro-conductive system! The global electrical circuit discharges in one area and recharges in another.

There are constant exchanges that maintain the delicate balance of the Greater Whole. Storms, cumulonimbus clouds, and other atmospheric phenomena are all part of this harmonious dance.

Regarding the appearance of electric charges in thunderstorm clouds, active and diverse research emerged throughout the **20th century.** Matteo Tavera, a landscape architect (DPLG - certified by the government) and founder of the Nature & Progress movement, offers an explanation in his work The Sacred Mission. He explains that large thunderstorm clouds, such as cumulonimbus, *constantly and generously* bombard the Earth with negative charges.

At any given moment on Earth, countless locations host these types of clouds, which play a role in recharging the Earth. They are not the only contributors; thunderstorms must also be considered in this continuous process of recharging the planet. Thunderstorms generate an inverted electric field compared to fair weather, also delivering negative charges to the ground.

TELLURISM

Electric fields exist in the ground; they are known as telluric currents. In his book The Sacred Mission, Matteo Tavera states: «There is a rather curious reason to mention why these currents have been studied more seriously. However weak they may be, they have a notable effect. Initially, they interfered with the oil exploration efforts of geophysicists by disrupting their operations. Unable to eliminate them, geophysicists had to adapt to their presence.»

These currents lie just a few centimeters below the Earth's surface and slightly above the ground. Though weak, they are essential for living organisms. They can be detected by creating two sufficiently distant poles, such as copper rods inserted into moist soil. By connecting an ammeter to a conductive wire linking the two poles, a current of a few milliamperes can be measured between the two buried rods. It is a weak current, but it exists. The composition of the soil, moisture, percentage of humus, and other factors influence these telluric flows and the soil's conductivity.

According to Matteo Tavera, there is also a natural current of electrons flowing from east to west, which fluctuates with the seasons, reaching its peak around **June 21.**

OPERATION

Atmosphere/soil exchanges

Point effect
Electro-osmosis
Electric tension
Soil
Electronegativity

Although it may seem complex at first glance, understanding electromagnetic ecology is essential to fully grasp how an aerial lightning rod–type antenna works. **Several distinct yet complementary effects are involved: air–soil exchanges, the point effect, the electric parasol effect, electro-osmosis, and electric potential.**

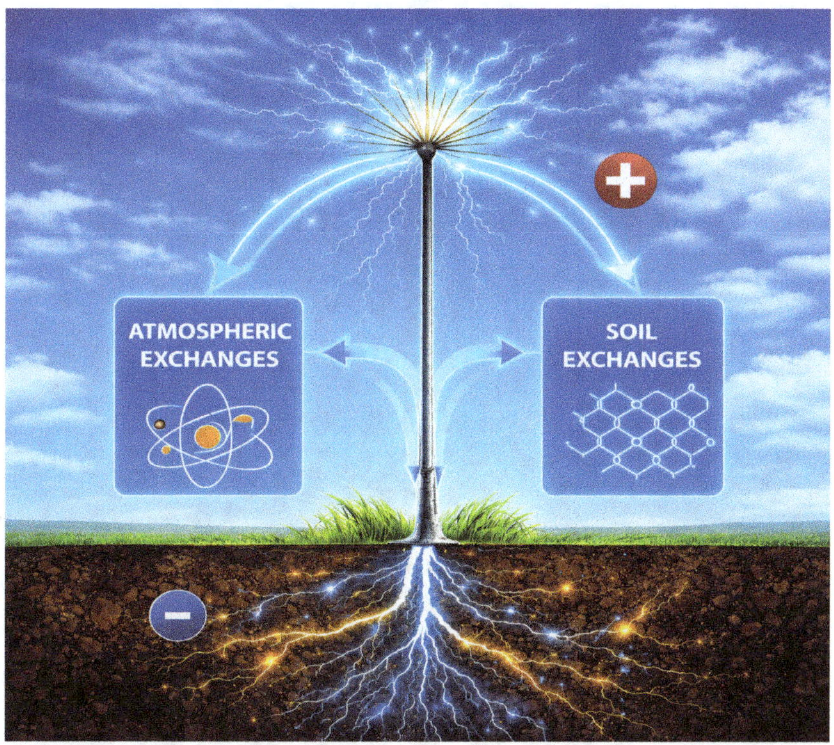

The Earth continuously releases negative charges. Electrons can rise along the antenna, escaping through its tips and dispersing into the air. The flow is therefore bidirectional: it can move from the ground to the air or from the air to the ground. The Earth acts both as a reservoir, releasing electrons, and as a sink, absorbing them. Due to its large mass, it can also replenish its charge. The antenna facilitates these exchanges, guiding electrons between the air and the ground. Additionally, the choice of metals can enhance these effects by creating battery-like interactions, further promoting the circulation of electrons.

Figure 23 : Electrons circulate between the sky and the Earth.

POINT EFFECT

Like a lightning rod, the tips of the antenna serve to maximize the exchange of electrons. The more tips there are, the more the interactions between the ground and the air (via the antenna) are amplified. Electrons can move back and forth through the antenna, using these tips as entry and exit points.

Figure 24 : The tips, entry and exit points for electrons.

ELECTRO-OSMOSIS

Moreover, water follows the movement of electrons, a process known as electro-osmosis. In dry soil, it is possible to draw water from the subsoil to the surface. To achieve this, it is advisable to bury the antenna deep enough to reach moisture, allowing it to bring up the precious liquid. The electron pump created by the antenna will stimulate the upward movement of water. Conversely, if your garden is too wet, it might be better to keep the installation closer to the surface.

Figure 25 : Water follows the electrons.

ELECTRIC TENSION OR POTENTIAL DIFFERENCE (IN VOLTS/FOOT)

The potential difference is essentially a difference in electron concentration between two environments. It is expressed in *volts* and is also referred to as *tension*. Simply put, the higher this value, the greater the difference in electron concentration between the two measured environments. A battery, for example, consists of two reservoirs: one rich in electrons and the other deficient; electrons then flow from the richer environment to the poorer one, generating an electric current.

> The greater the potential difference between the bottom and the top of the antenna, the more the electron pump can theoretically be stimulated. For optimal operation, the antenna should be placed at a minimum height of 21.3 to 23 feet. Below this height, the effects are still present but less significant.

Moreover, this phenomenon creates a more electronegative atmosphere in the surrounding air. Regarding these tension lines (in volts per foot), the antennas bend them, as schematically illustrated on the next page. Thus, it is not the same to have bare ground compared to ground with trees or a *lightning rod-type aerial antenna*! Their presence raises these lines, creating areas of lower tension around the metal masts, which in turn leads to environments richer in electrons.

Finally, at the top of the antenna, due to the deformation of these voltage lines, a significant potential difference (tension) is concentrated over a smaller height.

It is also interesting to note that electrons are natural antioxidants. For instance, walking in a forest rich in electrons can have positive health benefits.

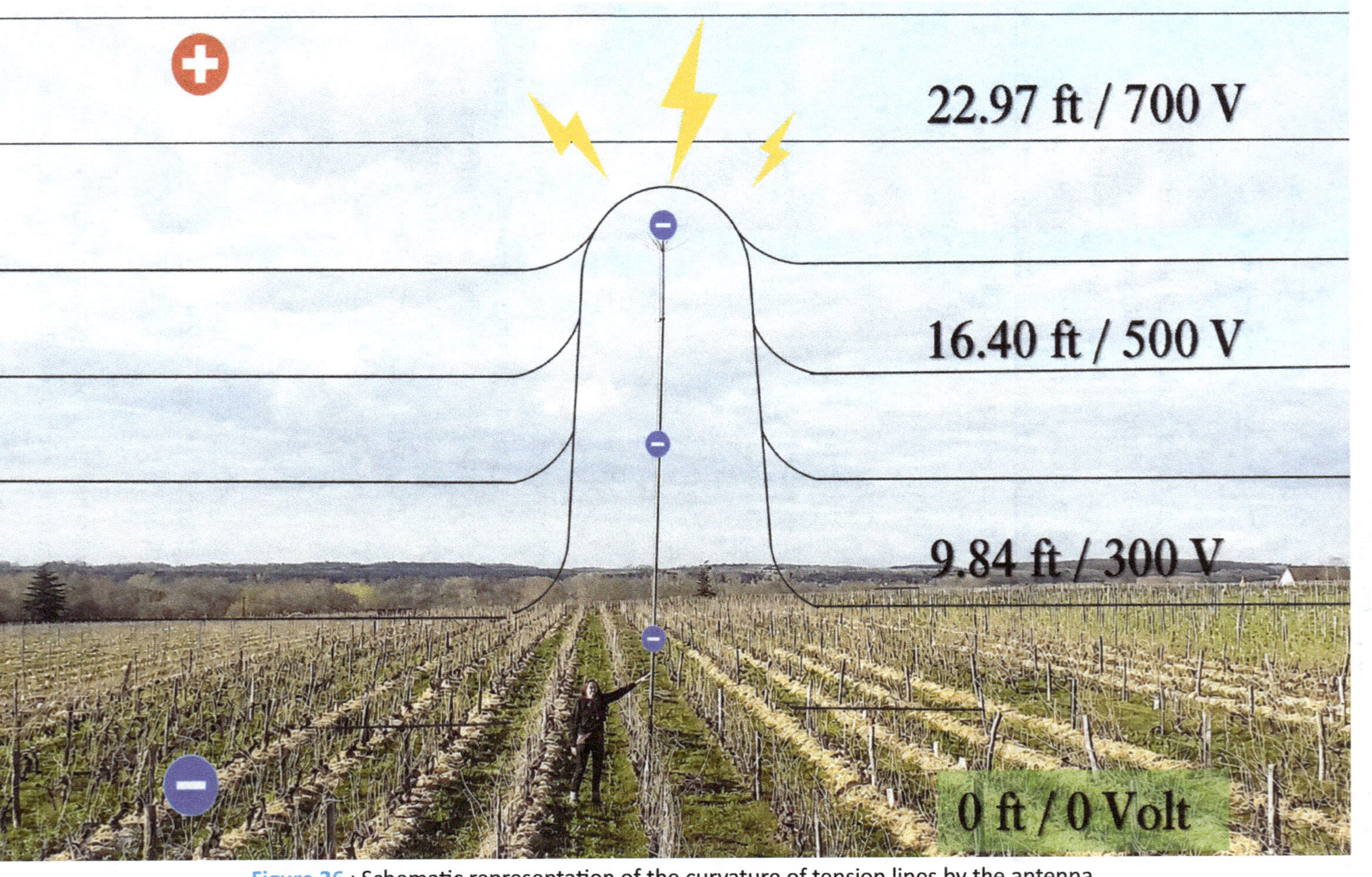

Figure 26 : Schematic representation of the curvature of tension lines by the antenna.

SOIL

Humidity, pH, humus content, clay percentage, texture, structure, the presence of paramagnetic materials (such as basalt), mulching... In fact, many factors influence the electromagnetic properties of the soil. These factors alter parameters such as conductivity, electron richness, and magnetic susceptibility.

As a result, it becomes clear that there is not just one type of context but countless variations.

Thus, **the bio-electromagnetic effects of antennas on plants can vary significantly, making their impact more or less noticeable.**

POTENTIAL DIFFERENCES, ELECTRONEGATIVITY, AND REDOX (OXIDATION-REDUCTION)

When constructing antennas, it is possible (but not mandatory) to use different metals arranged in a thoughtful way to take advantage of the natural electronegativity properties of metals. The goal is to promote a specific direction in the flow of electrons by creating a potential difference between the metals, thus optimizing the system.

Indeed, electronegativity varies depending on the type of metal. As a reminder, moving to the right and/or upward on the periodic table increases electronegativity (see the periodic table of elements using Pauling's electronegativity scale on the next page).

> **Electronegativity is the ability of a material to attract electrons. The higher the electronegativity value, the more effectively the material can capture electrons.**

By combining two metals with different electronegativities, a natural transfer of electrons will occur from the less electronegative metal to the more electronegative one, functioning like an electro-pump. It is sometimes said that a metal is electropositive, but in reality, electropositivity doesn't exist as an independent property. When a metal is described as electropositive, it is always in relation to another metal. For example, gold (Au) has an electronegativity of 2.54, while iron (Fe) has an electronegativity of 1.83; compared to gold, iron is considered electropositive because it attracts electrons less strongly than gold.

1	2	3	4	5	6	7	8	9	10	11	12	13	14	15	16	17	18
H 2,2																	He
Li 0,98	Be 1,57											B 2,04	C 2,55	N 3,04	O 3,44	F 3,98	Ne
Na 0,93	Mg 1,31											Al 1,61	Si 1,9	P 2,19	S 2,58	Cl 3,16	Ar
K 0,82	Ca 1	Sc 1,36	Ti 1,54	V 1,63	Cr 1,66	Mn 1,55	Fe 1,83	Co 1,88	Ni 1,91	Cu 1,9	Zn 1,65	Ga 1,81	Ge 2,01	As 2,18	Se 2,55	Br 2,96	Kr 3
Rb 0,82	Sr 0,95	Y 1,22	Zr 1,33	Nb 1,6	Mo 2,16	Tc 1,9	Ru 2,2	Rh 2,28	Pd 2,2	Ag 1,93	Cd 1,69	In 1,78	Sn 1,96	Sb 2,05	Te 2,1	I 2,66	Xe 2,6
Cs 0,79	Ba 0,89	Lu 1,27	Hf 1,3	Ta 1,5	W 2,36	Re 1,9	Os 2,2	Ir 2,2	Pt 2,28	Au 2,54	Hg 2	Tl 1,62	Pb 1,87	Bi 2,02	Po 2	At 2,2	Rn 2,2
Fr 0,7	Ra 0,9	Lr 1,3	Rf	Db	Sg	Bh	Hs	Mt	Ds	Rg	Cn	Nh	Fl	Mc	Lv	Ts	Og

La 1,1	Ce 1,12	Pr 1,13	Nd 1,14	Pm 1,13	Sm 1,17	Eu 1,2	Gd 1,2	Tb 1,2	Dy 1,22	Ho 1,23	Er 1,24	Tm 1,25	Yb 1,1
Ac 1,1	Th 1,3	Pa 1,5	U 1,38	Np 1,36	Pu 1,28	Am 1,13	Cm 1,28	Bk 1,3	Cf 1,3	Es 1,3	Fm 1,3	Md 1,3	No 1,3

Figure 27 : Periodic Table of Elements Using Pauling's Electronegativity Scale (source: Wikipedia).

**Decreasing order of
electronegativity:**

**Gold (Au): 2.54
Platinum (Pt): 2.28
Silver (Ag): 1.93
Copper (Cu): 1.90
Iron (Fe): 1.83
Zinc (Zn): 1.65
Aluminum (Al): 1.61**

The potential difference between metals is crucial in electroculture when selecting materials for the antenna, mast, or systems placed in or on the ground. This difference allows for the creation of varying battery effects and electron exchanges.

However, be mindful of the specificities of alloys (brass = copper + zinc), as they also have their own unique electronegativities.

Self-build

Alone or in combination ?
Construction details
Installation of the top
Risk of lightning ?

WARNING : Not all systems are created equal.

Some methods are more or less effective, and there is no «miracle recipe» that can be applied universally. There are paths for research, reflection, and implementation to explore. The goal is to move forward together toward positive results and increasingly effective antennas tailored to different contexts and objectives. In fact, there could almost be as many antennas as there are electroculturists...

My aim here is to contribute, modestly, by guiding you along the path; to (re)discover examples that will help you make your own choices and carry out your own experiments.

THE AERIAL ANTENNA SYSTEM HAS SEVERAL VARIATIONS:

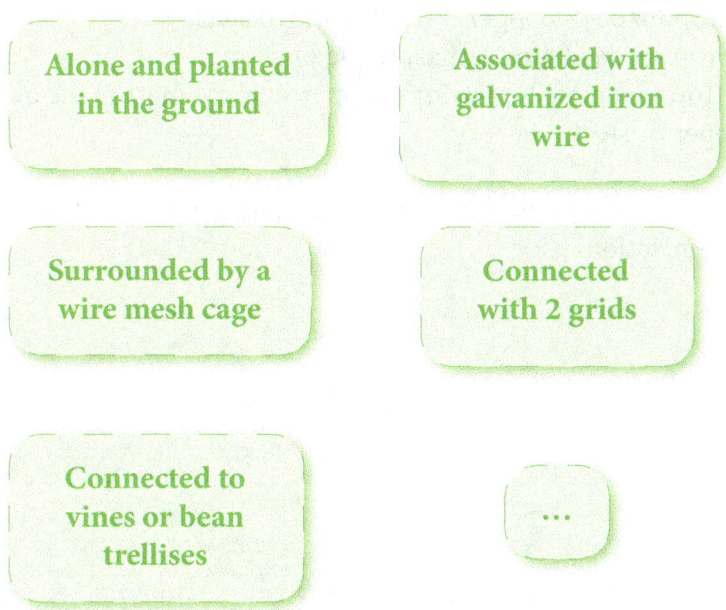

Alone or in combination?

1/Alone

The taller the antenna, the greater the potential difference between the top and the bottom, and the more significant the results will be. The optimal height for an effective antenna is at least 21.3 feet, though it can range from 6.6 to 23 feet (or even higher). When positioned lower, the effects will still be present but diminished.

The area of influence around the antenna is approximately equal to its height. This will stimulate air-soil and soil-air exchanges at the base and around the antenna. However, the presence of taller trees nearby may interfere with its function or cause it to perform poorly.

Simple choice of metals : (non-exhaustive)

A/ Top made of aluminum + mast made of copper.
B/ Top made of copper + mast made of steel.
C/ Top made of zinc + mast made of copper.
D/ Top made of aluminum - copper - zinc + mast made of copper or steel.

Ideal for : vegetable gardens, permaculture mounds, water dynamization...

The copper mast can be replaced by a wooden post. In this case, the top of the antenna will be connected to the ground using either copper electrician's cable or galvanized wire; the former is highly conductive, while the latter is ferromagnetic.

In the photo below, there is a very simple antenna: a concrete reinforcing bar made of 0.31-inch steel, topped with zinc tips from a gutter.

Figure 28 : Loïc in his garden with a simple homemade antenna.

2/Surrounded by a wire mesh cage

The antenna will be planted in the ground and surrounded by a wire mesh buried at depths of 8, 12, 16, and 20 inches, which may protrude above the surface or remain below. This system can be ideal, for example, for raised garden beds or squares, with the wire mesh placed near the edge of the bed, as seen in elevated gardens.

Figure 29 : Antenna surrounded by a wire mesh cage in a wooden bed.

In fact, many people already use wire mesh in garden beds to deter rodents, so why not for electroculture? The goal isn't necessarily to place the wire mesh directly in the middle of the garden. Practicality must be a key consideration during installation!

The distance between the wire mesh and the antenna should be no greater than the height of the antenna. The wire mesh cage should not have a bottom or a top; its purpose is solely to facilitate the exchange between the antenna and the mesh, and vice versa. This system can be implemented in a pot or directly in the garden (if using a pot, ensure the substrate can be connected «*to the ground*»).

> It is important to avoid placing the wire mesh directly against the edge of the wooden bed. Ideally, there should be soil all around it. The natural current will flow multidirectionally, both in the soil and above ground. This will amplify the exchange of electrons between the antenna, the wire mesh, the soil, and the air. In short, energy will circulate...

There will be more electron circulation in the electrocultivated area compared to an unequipped plot, generally allowing plants to benefit from these exchanges.

Simple choice of metals : (non-exhaustive)

A/ Top made of aluminum + mast made of copper + galvanized iron wire.

B/ Top made of copper + mast made of steel + galvanized iron wire.

C/ Top made of zinc + mast made of copper + galvanized iron wire.

D/ Top made of aluminum - copper - zinc + mast made of copper or steel + galvanized iron wire.

Ideal for : square gardens, planters, raised beds...

3/In connection with two grids

This system is designed to create a potential difference and enhance the natural micro-currents between the grids. Ideal for use in vegetable or market gardening beds, it consists of an antenna isolated from the ground and topped with spikes. A copper wire, for example, will connect the antenna to the first buried grid. This grid should measure several meters in length (depending on the cultivated area) and will have a variable height. It is important to ensure that the mast (or conductive wire) connecting the top to the grid is well insulated from the ground so that the electrons captured from the air flow to the first grid (and not directly into the soil). This grid will be buried at a depth of several tens of centimeters.

> The electrons accumulated in the first grid will tend to be attracted to the second one, with both grids aligned north-south along the axis of the Earth's magnetic field. This setup will create a stimulation zone between the two grids.

Furthermore, the grids in the ground will be positioned perpendicular to the east-west axis. This is because a telluric current flows from east to west, which would contribute to the transfer of electrons from the grid located in the east to the one in the west.

Simple choice of metals : (non-exhaustive)

A/ Top made of aluminum + mast made of copper + galvanized iron grids.
B/ Top made of copper + mast made of steel + galvanized iron grids.
C/ Top made of zinc + mast made of copper + galvanized iron grids.
D/ Top made of aluminum - copper - zinc + mast made of copper or steel + galvanized iron grids.

Ideal for : market gardening beds, square gardens...

Figure 30 : Cross-section of a cultivated bed with its two grids and the aerial antenna.

4/Connected to a galvanized iron wire

The galvanized iron wire extends the range of the antenna along its entire length. This wire can measure several meters, or even tens or hundreds of meters, as long as it is properly aligned with the Earth's magnetic field using a compass. Once aligned, the wire will become magnetized. The antenna is placed on the geographic south side, while the wire extends toward geographic north. This setup allows the Earth's magnetic field to pass through the entire system, first through the antenna and then along the wire, as terrestrial magnetism moves from south to north.

The wire will be placed on the surface or buried, in contact with the root hair zone, at a maximum depth of 19.7 inches. Being ferromagnetic, it will become magnetized under the influence of the Earth's magnetic field.

Additionally, the iron wire to which the antenna is connected can be aerial, meaning it will not be placed in or on the ground, as seen in vineyard plots, for example. The advantage here is that the metal structure already exists, and all that remains is to connect the antenna.

It is crucial to ensure that the antenna is properly connected to the main conductive wire of the vine row. The stakes supporting the grapevines should be made of metal, connected to the main wire, anchored at the base of each vine plant, and extending above the main galvanized iron wire (preferably—**see the schematic on page 86)**.

If the wire is oriented along the magnetic field's axis, an additional magnetization effect occurs. However, even without this alignment, there is still an effect due to the electron exchanges between the top of the antenna, the ground, and through the wire and metal stakes.

This setup will stimulate the exchange of electrons between the air and the soil, as well as between the soil and the air, at the base and around the antenna, along the wire, and approximately 3.3 feet (1 meter) on either side. If the wire is not oriented parallel to the magnetic field, beneficial electron exchanges will still occur; however, the effect of the Earth's magnetism will be minimal or nonexistent.

Figure 31 : Galvanized iron wire buried in the ground over small fruit trees.

Figure 31 : Aerial antenna on the row of vines.

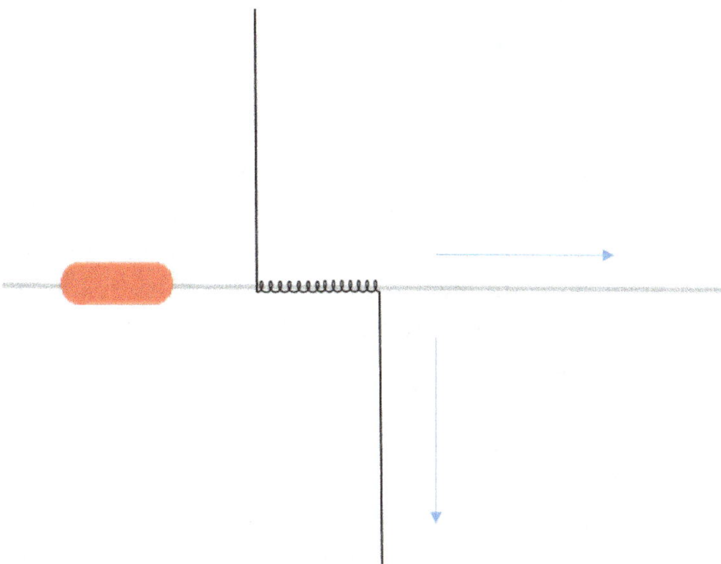

Figure 32 : Illustration of the method for attaching the tension wire to the main wires when using a suspended wire for the vines.

In his **20th-century** writings, Justin Étienne Christofleau suggests placing a porcelain insulator on the wire of the vine row. This insulator should be positioned just after the first strut post, located on the geographic south side. The wire from the antenna should then be connected just after this insulator (see schematic on the previous page). According to him, this setup ensures that the natural current is not lost by being absorbed into the decomposition of the first post.

Simple choice of metals : (non-exhaustive)

A/ Top made of aluminum + mast made of copper + galvanized iron wire.
B/ Top made of copper + mast made of steel + galvanized iron wire.
C/ Top made of zinc + mast made of copper + galvanized iron wire.
D/ Top made of aluminum - copper - zinc + mast made of copper or steel + galvanized iron wire.

Ideal for : This system can also be applied to structures that already have an existing galvanized iron wire, such as those used for kiwis, vines, or small fruits. However, care must be taken not to obstruct the path of the tractor!

5/ Structure for climbing beans or others

The antenna is simply connected to the central part of the mesh, similar to the geomagnetifier used by Brother Paulin in the 19th century (see page 26 of this book). His system was directly planted in the middle of the mesh structure in the ground and did not align with the Earth's magnetic field. In this case, it is necessary to imitate this concept, but by positioning the metal system above the ground.

Additionally, it might be worthwhile to experiment by planting metal stakes at regular intervals. These stakes will connect the plant support mesh to their roots (and thus to the top of the antenna!). Lastly, it is crucial to ensure that the antenna is properly connected to the mesh via a conductive wire.

Figure 33 : Antenna placed in the middle of the climbing beans.

DETAILS OF THE CONSTRUCTION OF THE UPPER PART

The antenna must be equipped with numerous tips to enhance the point effect and improve energy exchanges between the atmosphere and the ground. It can be made from a single metal and constructed quite simply, or it can be more complex, consisting of different metals (or even more materials).

MULTIPLE RODS INSERTED INTO A COPPER PIPE

The tips can be made from galvanized steel, copper, aluminum, or zinc. It is essential to ensure good contact between the metals, and welding may be useful to achieve optimal results (or at least to ensure proper electrical contact between all the metals!). Additionally, special attention should be given to iron, as it rusts quickly if not galvanized.

IMPROVEMENTS

In the two photos on the next page, a computer radiator made of aluminum is placed in contact with the main copper mast and positioned perpendicular to it. The goal is to increase the surface area in contact with the atmosphere and to extract electrons from water droplets. One advantage of aluminum is its lightness compared to other metals. The sharp needle, approximately thirty centimeters long, facing the radiator, should ideally be ferromagnetic and oriented toward geographic south (aligned with the Earth's natural magnetic field). This needle serves to capture magnetism and enhance the effects of the aerial antenna (see also Figure 13 at the beginning of the book for reference).

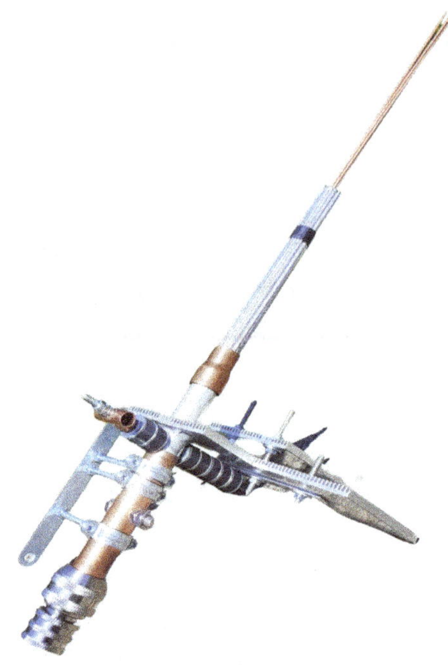

Figures 34 et 35 : An improved and effective antenna in construction.

Moreover, I believe that taking the time to appreciate the beauty of a creation is already a step toward harvesting the heart, which is, in itself, a form of harvest!

> These examples clearly demonstrate that there are no limits to creativity, knowledge, or imagination. As a result, one can design antennas that are both functional and artistic...

Additionally, it is entirely possible to stay within the virtuous cycle of reuse and repurposing of waste, known as the *circular economy*. By recovering materials, we can effectively limit waste.

Figure 36 : Recovery at the scrap yard, metals are sorted by type.

RISK OF LIGHTNING?

Of course, I strongly advise against placing a system on the roof of your home or any other location that requires protection from lightning; this is not a lightning rod! Although I have personally taken measurements on an aerial antenna during stormy weather, I highly recommend avoiding them during thunderstorms. Trees and any natural or artificial elevated points can potentially attract lightning. Additionally, the higher and more exposed the area, the greater the risk of being struck!

To date, since 2012, I have not received any reports of antennas being struck by lightning. And if it does happen... isn't it better for lightning to strike the electroculture system rather than your house or a beautiful tree? After all, don't you already have TV antennas on the roof of your house? Aren't there already numerous natural points in your garden or on your street?

One only needs to observe to realize it...

INSTALLATION OF THE TOP OF THE ANTENNA?

The antenna should be at least 6.5 feet (2 meters) above the ground and connected either to the ground or to a mesh system located inside or on the ground.

In the **1910s and 1920s,** Justin Christofleau positioned them at heights exceeding 19.7 feet (6.5 meters), which I believe would be the minimum height required to achieve good results. Generally, the higher the antenna is positioned, the more significant the visible effects. This increased height allows access to higher voltages. Some masts were even erected over 39.4 feet (12 meters) in the past, and up to 65.6 feet (20 meters)! However, I have also observed effects from antennas positioned lower than this, though they are likely diminished.

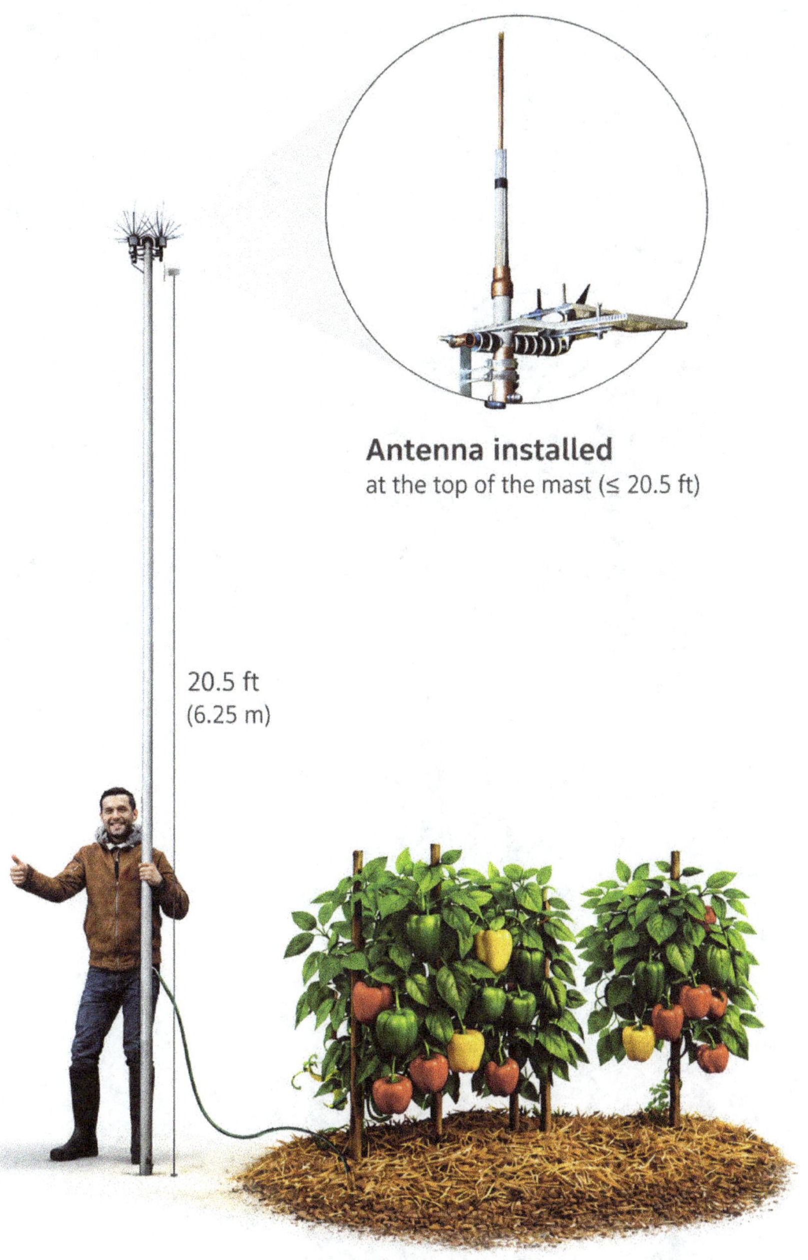

Antenna installed
at the top of the mast (≤ 20.5 ft)

Can be installed on wood, metal, or concrete structures. If installed on a wooden post, the antenna must be properly grounded using a conductive wire (galvanized steel, copper,

Testimonials

A promising 21st century

THE 21ST CENTURY AND CONCRETE MEASURES

Tension : It is expressed in *volts* and measures the difference in electron concentration between two environments. In other words, it represents a difference in electric potential.

First example : To illustrate, let's imagine that the marbles here represent electrons. Visually, it is clear that there are more electrons on the right than on the left. Since tension is always measured between two points, one lead of the voltmeter should be placed on the left and the other on the right. A result will appear on the screen in **volts**, for example, 100 millivolts. This measurement indicates the difference in electron concentration between these two environments.

Figure 37 : Tension is a difference in electron wealth between two environments.

> *With electroculture antennas, the voltage measurements are generally low, typically in the range of millivolts or volts.*

Second example: In this case, there are even more marbles on the right than in the previous example. Therefore, the difference in electron concentration between the two environments is greater. As a result, the voltage value will be higher than the previous measurement of 100 millivolts.

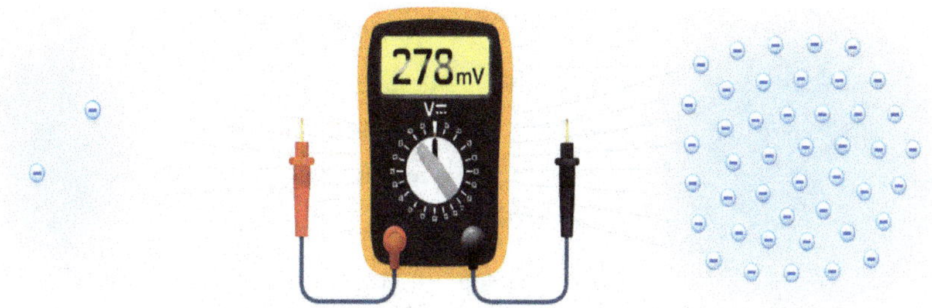

Figure 38 : The higher the value, the greater the difference in electron wealth between the two environments.

Next, **amperage**: it refers to the **current intensity**. This occurs when electrons flow between two measurement points in a certain quantity. In the example below, the marbles (electrons) from the richer environment move toward the one that is less rich.

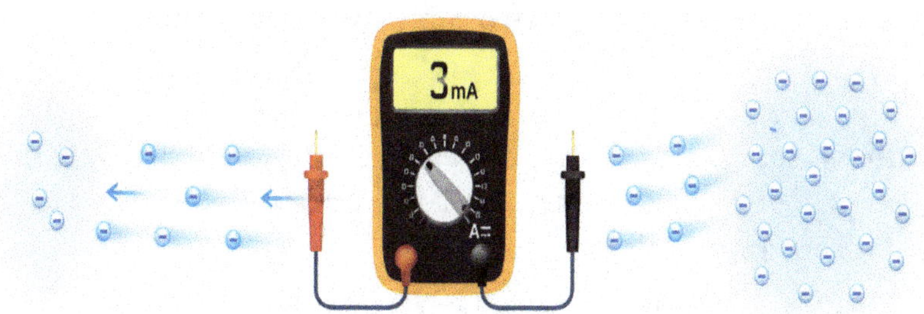

Figure 39 : Circulation of electrons from right to left.

In this final example, there is a current flow measured in *amperes.* The higher its value, the stronger the current intensity, and the greater the number of electrons passing through.

In electroculture antennas, the current flow is measured in milliamperes because the currents are weak and continuous.

Why am I sharing this with you? One might think that we are straying from gardening and agriculture. In fact, my goal here is to demonstrate that these techniques are measurable and concrete. This is not a deception!

In **2022**, with Gilles and Michel, two of my retired electrician neighbors, we embarked on the construction of a mega antenna. My initial hypothesis was that by building a large antenna, I would have a better chance of measuring electron exchanges than with a smaller one. I wanted to validate, through measurements, the information I had gathered from my historical readings.

A question arises: do antennas promote the circulation of electrons between the sky and the earth? How can this be measured concretely?

We built a device using a large copper pipe with a diameter of 0.79 inch and a height of 16.4 feet (5 meters). The top was equipped with numerous tips made mostly of aluminum.

Indeed, aluminum is lightweight and has a significant electric potential difference compared to the copper mast. We also equipped it with numerous tips made of zinc and copper. Additionally, we used a large radiator purchased from Emmaüs, designed to act as an electron collector, extracting electrons from rain droplets.

I won't dive into all the technical details of this construction here. However, I encourage you to watch the explanatory video on YouTube titled «*Construction mega antenne + preuves échanges électrons*» by *YouTube Permafutur* for clearer and more comprehensive information.

Figure 40 : Loïc in the process of constructing an aerial antenna.

Once the construction was complete and the installation was in place, the top of the antenna reached nearly 26.24 feet!

Subsequently, I was able to take numerous voltage (volts) and current (amperes) measurements over several months.

It turns out that between my antenna and the mesh system I had installed in the ground, there was an average of 1.5 milliamperes of regular direct current. So, there is indeed current circulation! During stormy periods, I even measured intensity peaks and values that were much higher!

Figure 41 : 1.71 milliamperes of current between the antenna and the mesh in the ground.

Figure 42 (previous page) : The aerial antenna and a huge lettuce.

At the time of installation, we measured 800 millivolts directly between the antenna and the ground. In one instance, I was able to exceed one volt, while at other times, the voltage dropped below 500 millivolts. Everything charges and discharges, and the electrical nature fluctuates just like the weather. Therefore, these measurements are provided as a rough guide and are not exhaustive.

They will indeed vary based on factors such as the height of the antenna, its materials and characteristics, the day of the measurement, the weather, the presence of metal structures on the ground, and more.

In any case, this testimony validates the basic theory: there are indeed electrons circulating between the antenna and the ground.

Figure 43 :
Construction of the
«mega» antenna.

USER TESTIMONIALS

Alain Deharbe : His Research and Experiments

Discovery of Electroculture

«*It was in the 2000s that I first became acquainted with energies capable of boosting our vegetable crops. My initial approach involved simply planting a few 6.56 feet steel rods at the base of my favorite plants and using metal trellises as supports, yielding results that I already found surprising. In 2010, I came across the book The Secret of the Patriarchs by Marcel Violet, in collaboration with Michel Rémy and Christian Beau. From that moment, I was determined to explore this vast field further. Armed with a hazel wand, I discovered two water veins, one and a half meters deep, which contributed to feeding the many ponds located about 547 yards away and downhill. During the construction of a 114 feet² greenhouse, I decided to fully integrate electroculture and telluric energies by building it on this energy point.*»

The Essential Tool for Capturing Energies: The Aerial Antenna

«*To harness this energy, I was inspired by Justin Christofleau's antenna and built mine using modern materials to guide the energies to the planting beds through a network of copper wires. The first version of this antenna featured two aluminum tips—one pointing toward the sky and the other toward the south—along with six zinc tips sticking out in all directions, and two copper tips oriented east/west. Additionally, an aluminum radiator was included, reacting to the temperature difference between its two sides, similar to when receiving raindrops that carry electrons, acting 'like a voltaic pile,' based on Christofleau's antenna design from his 1927 book Electroculture (or see page 41 here).*»

«My choice to use aluminum is based on its electron mass of -1.66, zinc at -0.763, and copper at +0.337, rather than the steel used by Christofleau, which is at -0.44. This choice aims to achieve a maximum potential difference with the surrounding air at approximately 6 meters above the ground. (It should be noted that Christofleau likely did not use aluminum due to its high cost, which only became more affordable in France starting in 1930, thanks to a new manufacturing process.) The entire setup is wound with a coated copper wire, one end in contact with the tips and the other with the copper mast to which the cable is connected. This creates a sort of coil that induces a magnetic field through the flow of electrons, which is then transmitted to the mast. After adding this, I observed an additional 30 millivolts.»

«When measuring the potential difference between the antenna cable and the ground of the planting beds, it averaged around 380 millivolts during installation in 2013 on clear days, easily rising to over 500 millivolts during rainy weather. While plants benefit from this input, it is not the only beneficial energy. What I cannot measure are the electromagnetic waves, particularly Schumann waves, which are transmitted and provide additional benefits. The electrical current, however weak, acts like a carrier current, ensuring a measurable potential capable of being transmitted over a significant distance. In this way, 12.67 square yards (yd²) of greenhouse and two rows of 19.7 feet in the outdoor garden benefit from this energy generated by my single antenna. Even better, there is enough energy left to dynamize water...»

Figure 44 (previous page) : Aerial antenna of Alain Deharbe.

A somewhat high-tech antenna that appeals to birds.

Figure 45 : More complex optimizations on Alain's antenna.

*«Over the years, I have noticed that the potential difference has decreased. By **2017**, four years later, it had already dropped by nearly 100 millivolts. Despite cleaning the contacts, simplifying the entire network, and refurbishing the antenna, it is slowly but surely degrading. I cannot say whether this is a general phenomenon or specific to this installation; however, nearby trees are growing and may also be contributing to the blockage.»*

«In mid-2020, I decided to reinvigorate my antenna to improve my vegetable crops, making several modifications. I had noticed that what was lacking were flexible tips that could vibrate with the slightest breeze, so I added a bouquet of galvanized steel tips, oriented in all directions. Having observed that a coil could increase the potential difference, I added a new one overlapping the first, into which I introduced a nearly constant current. For this, I used a small 1.2-volt photovoltaic panel at 40 mAh to power a quartz buzzer.»

«To keep my 'consumable' energy as natural as possible, this circuit is independent and completely isolated from the network that nourishes my plants. The negative at the bottom and the positive at the top serve only to generate an electromagnetic field, which tends to accelerate the flow while drawing from the field generated at its center by the Ighina spiral. This quartz has the unique ability to vibrate across a wide frequency range, varying according to the current intensity, which comes from the sun powering the photovoltaic panel. This principle aligns with Marcel Violet's method of water dynamization in its mechanical version, covering a broad spectrum of waves to inevitably capture the range of Schumann waves. The quartz, attached to the mast, has a very stimulating effect on the birds, which now come in greater numbers and build their nests on the nearest one. This becomes a bit problematic as they start entering the greenhouse, pecking at the tomatoes, and even trying to build nests inside...

With this setup, my antenna has regained its initial capacity, now showing 380 millivolts in the greenhouse and 150 millivolts at the end of the 98.4 feet wiring leading to my outdoor vegetable garden. In my opinion, this demonstrates the ability to transmit beneficial information captured by the antenna over long distances, without needing additional procedures like magnets, cones, or even multiple antennas. My potatoes will surely yield the spectacular results I observed until 2017, with at least 3.97 pounds per plant.»

«The natural voltage of plants can be measured with a simple voltmeter, but it varies depending on the plant species and their sizes. In tests on tomato plants about 19.7 inches tall, the average voltage was around 15 to 25 millivolts (mV); after the changes made to the antenna, it increased to 35 to 45 millivolts (mV). The electrons are stimulated, the cells become more vigorous, and the plant is in better health. Each electron carries electromagnetic information. This leads to more effective disease resistance, significantly reduced water consumption, and even benefits for the soil fauna.»

Testimony from Yann Gaulupeau about his electrocultivated vineyard plot.

Yann explains: «This is an old vineyard, 40 years old, with Chenin grafted onto 3309 rootstock. The plot measures 3,588 square yards (yd²), but with a 40% deficit, only about 2,392 square yards (yd²) is actively cultivated. At its peak, the vineyard used to yield 700 liters, but last year, the production dropped to just 200 liters.

In 2022, I decided to implement a basic aerial antenna, which I carefully connected to the fruiting wires. It is not connected to just one row of vines but to all the rows (oriented east-west)! Yes, one antenna for my entire vineyard plot! The 23-feet mast is positioned in the east, with its connecting wire oriented north-south, perpendicular to the rows of vines. All the rows are thus connected to the aerial antenna in the east via this main wire (see schematic on page 110). This technique is possible here because we do not use a tractor for treatments; otherwise, the aerial antenna's wire would be somewhat obstructive. Additionally, it's worth noting that the vineyard is managed using biodynamic permaculture and Conservation Agriculture practices.»

The amendments: *manure and straw under the ridge, followed by a mixture of PP500 and 100 diluted eggs.*

The treatments: *no sulfur, no copper, only fermented extracts and herbal teas. Nine applications of nettle and comfrey extracts, plus horsetail teas, with two spring applications including honey. There was a bit of mildew on* **June 30,** *but it was quickly stopped by the horsetail; overall, it was an easy year.*

Of course, the electroculture antenna is not the sole factor; the year was good as well... but around my small vineyard, many winemakers had different results. I am located at the center of a plateau, which has allowed me to compare my practices with those of my neighbors. In 2022, we had so many clusters that, fearing they wouldn't ripen evenly, we removed some! This was also to comply with the AOC limits! Additionally, some clusters were not fully ripe at harvest.

Now, some numbers: from these 1,000 vines on this plot, we harvested about 2.5 tonnes of grapes, which yielded around 1,600 liters or 2,000 bottles from 2,392 square yards (yd²) of vines. That's one bottle per square meter! Two bottles per vine! Clusters weighing 2.2 pounds, and up to 30 clusters per vine! For a super quality juice at 53.6°F! Noble botrytis almost everywhere!

As for electroculture, it doesn't do everything, but it helps amplify the exchanges between ions and cations in the soil. In my opinion, it works particularly well in non-tilled soil. I suspect it's less effective in tilled or bare soil (even in organic farming!), as such soil no longer has well-structured horizons.

For me, electroculture likely accounts for 20% of the final result.»

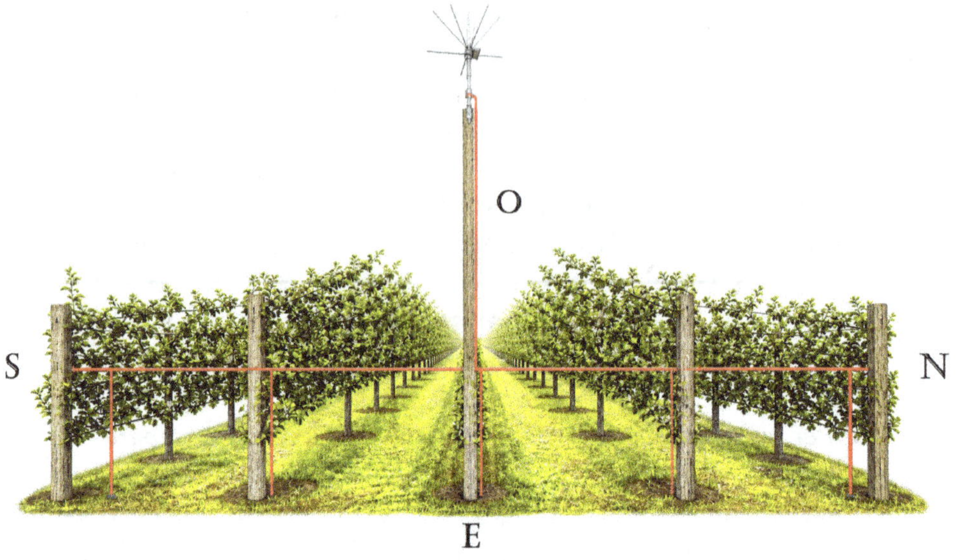

Figures 46, 47 et 48 : Schematic of the aerial antenna installation, beautiful grape harvest, and photo of the plot.

Dynamized water : promising research?

Information on Water
Active Electrification
Rain
Testimonials

This topic is very broad, and it would be reductive, even risky, to cover it in just a few pages. However, here are some facts and ideas for those who wish to explore its possible applications in the field.

INFORMATION ON WATER

For some, water is seen as a computer capable of storing both physical particles and energetic information, which it later transmits to the living or non-living elements it interacts with.

This could be partially explained by its electromagnetic properties: «...The positively charged hydrogen atoms in water molecules attract the negative ions of the substances they touch, while the oxygen atom, carrying a double negative charge, provides its positive ions to maintain balance. This allows water to break down substances into their constituent elements, capturing oxygen, nitrogen, and carbon dioxide from the air, as well as calcium, potassium, sodium, and manganese from rocks,» according to Alick Bartholomew in *The Genius of Viktor Schauberger*. However, this does not explain everything.

Biodynamics and homeopathy utilize this information transfer property. For instance, Rudolf Steiner, a pioneer of biodynamic agriculture, recommended that during biodynamic preparations (BD 500, BD 501, etc.), a small amount of the preparation be diluted in a large quantity of water. The material used is minuscule. The liquid must be stirred in a large bucket for an hour to integrate the information from the natural additive into the water. This stirring process, which involves creating vortices in both directions, forms an ellipsoidal shape (like an egg), followed by chaos, before returning to the vortex. From an electromagnetic perspective, when a vortex forms in water, it creates a potential difference between the top and bottom of the cone, subjecting the water to a mild electrical treatment during stirring.

After diluting the preparation and completing the stirring process, the water is sprayed onto the plants. Although only traces of the original mixture remain, the beneficial effects on the plants are still visible. Based on this principle, the information from the preparation is transferred to the water, which then transmits it to the plants. The plants, in turn, recognize and respond to this information.

Biodynamics, an organic farming method developed by Rudolf Steiner, aims to harmonize agricultural practices with cosmic and natural rhythms. It encourages performing these tasks with positive intentions, suggesting that our emotions and intentions can influence the vitality of water. This concept was further explored by scientist Masaru Emoto in *The Hidden Messages in Water*. Although these ideas remain controversial and require further scientific validation, I have personally encountered many passionate gardeners and agricultural professionals who attest to the success of these practices, including winemakers and market gardeners.

Once again, **I believe it is essential to maintain a holistic perspective while rigorously honing critical thinking— verifying sources, cross-referencing information, and most importantly, experimenting and analyzing—in order to continuously strive toward more sustainable fertility!**

ACTIVE ELECTRIFICATION OF WATER

Extraordinary experiments were already being conducted in the **18th century,** a time when the domestication of electricity was underway. Abbé Jean Nollet, one of the pioneers of electroculture, was very active in this field during his time. He was also a member of the Academy of Sciences and taught experimental physics.

Figure 49 : Abbé Bertholon sprays electrified water on trees.

In one of his experiments, Abbé Jean Nollet electrified water to mimic the properties of storm water, which is more electrified than regular water. He built a tank through which he passed an electric current. From the tank, a hose connected to a manual pump allowed him to spray the water. To insulate himself from the ground, Abbé Nollet stood on a small wooden table coated with beeswax. He then sprayed the plants and fruit trees. According to writings from that period, this method improved the growth, development, and disease resistance of the plants.

See the book *De l'électricité des végétaux*, which discusses the effects of atmospheric electricity on plants, its impact on plant health, and the medical and nutritional-electric virtues.

RAIN

At the beginning of the **20th century,** Simpson George developed an instrument to measure the charge of raindrops, which he discusses in *The Electricity of Rain and Its Origin in Thunderstorms.* I found this information in the thesis *Effects of Various Electrical Fields on Seed Germination,* presented at the University of Iowa in **1968. During 1907-1908**, an investigation was conducted at the Meteorological Office of the Indian Government in Simla. Systematic recordings were made using suitable instruments to measure the electricity carried by rain throughout nearly the entire rainy season. He summarized his data as follows: «*71% of the total time that charged rain fell, it was positive; 75% of the electricity brought by the rain was positive; most of the light and regular rains have a negative charge, while heavy rains almost always carry a positive charge.*»

TESTIMONIALS FROM ALAIN DEHARBE IN 2022

Passive Dynamization of Water

«*According to the principle developed by Marcel Violet and Stanislas Bignand, water dynamization is achieved by capturing the subtle energies and memories stored in the water after prolonged contact.*»

«It is from this idea that, following the same principle as for my crops, I applied this energy to a 5-liter container intended for watering seedlings, using two copper strips—one connected to the antenna and the other to the ground. A potential difference of 100 millivolts can be measured between the two electrodes submerged in the container; thus, it is not possible to speak of electrolysis, which only occurs from 1.23 volts. To ensure that the water thoroughly absorbs the information it receives, I leave it for about fifteen days—except after a storm, when, from experience, I know it is much more charged than usual. Dynamizing water from an electroculture antenna is, in fact, as easy as connecting a bulb to a battery...»

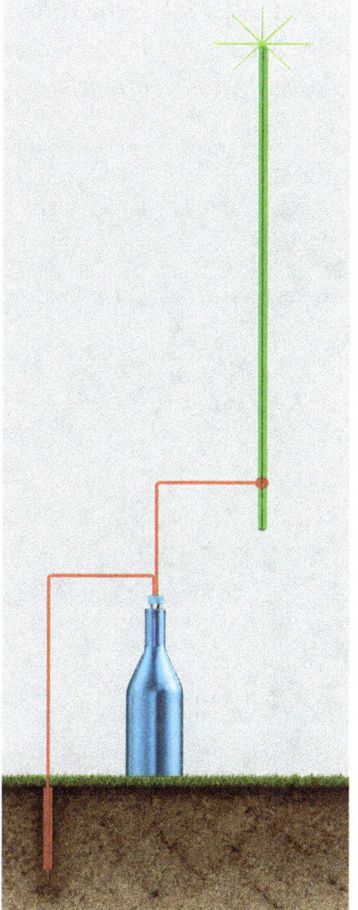

Figure 50 : One strip of copper connected to the antenna and the other to the ground.

Figure 51 : Connection of the antenna (in green) to the water bottle, with a ground connection (in red).

Experimentation on Seedlings

«Following my usual practice, each year, I soak my largest seeds for 24 hours in dynamized water before planting; from the very first season, I noticed faster and more vigorous emergence.»

«For comparison, I conducted a small experiment to demonstrate this vigor enhancement. I filled my container with tap water, allowed it to dynamize for about 40 days, and then used 6 seeds of a Maxima squash variety, 'Miniature Green Hubbard,' from the same batch.

I soaked 4 seeds in dynamized water and 2 seeds in non-dynamized water that I had let rest for 15 days. In both cases, the water came from the tap. The next day, I planted them and watered everything with the water from their soaking. Eight days later, the difference was visible, even though I accidentally used dynamized water for misting all the seedlings, forgetting about my experiment!»

Figure 52 : At the back, the seeds soaked in dynamized water, and at the front, the controls.

«Since I noticed a 'boost' effect when spraying my seedlings, I decided to take the concept further and spray this water on my crops. I observed wider leaves and a more pronounced green color. Year after year, as my confidence in the virtues of this dynamized water grew, I found myself spraying it on a tomato plant outside my greenhouse that was afflicted with mildew. The improbable happened: the progression of the disease stopped. I must admit that since then, I regularly use it on my entire vegetable garden, from planting until the end of the season!»

Water for the Aquarium

«I equipped a glass bottle with two carbon electrodes instead of gold, which is the most neutral yet conductive element, and treated my aquarium using the same method because it was causing me problems.»

Figure 53 : Bottle connected to the aerial antenna.

«One glass per month for six months, and my aquarium turned into a true jungle, requiring weekly pruning. No more buying plants for the aquarium! This was enough to significantly reduce the mortality rate of the platys and greatly improve their reproduction: the population multiplied tenfold, nearly reaching unmanageable levels. Of course, this should be used sparingly. My indoor plants also enjoy being sprayed. Even though they aren't placed in direct southern exposure and receive very little light, they are thriving beautifully.»

An Inexplicable Peculiarity

«It may be 25.7°F (−3.5°C) in the greenhouse, but the water does not freeze. Even at 20.3°F (−6.5°C) at the end of **January 2018**, and this year, when everything was frozen and my bottle was resting on a concrete block... It seems difficult to attribute this to the low electric charge; if that were the case, the industry would be using it. Clearly, another phenomenon is at play...»

Antenna in the Watering Tank

Mathieu Cailleaud shares a photo on **the next page** of one of his electroculture installations. He has been practicing for several years and attests to observing positive results in his vegetable garden and on his hemp crops.

In this setup, he has connected his aerial antenna directly to the water tank. The antenna rises to a height of 16.4 feet (5 meters) and is insulated from the ground by its mast. An electrical cable runs along the mast from the top and plunges into the water. A second cable serves as a ground connection and exits the tank. This cable is buried 11.8 inches (30 cm) deep, down to the moisture layer. It is crucial not to forget to include the ground connection.

Figure 54 :
Water tank connected
to an aerial antenna.
≈ 11.8 inches

MAGNETS AND MAGNETIC TREATMENTS

History
Electromagnetic Ecology
Scientific Studies
Description and Use
Implementation
Testimonial

HISTORY AND ELECTROMAGNETIC ECOLOGY

The term «magnetoculture» is sometimes used to refer to practices that utilize magnets to interact with plants. This terminology was first employed in the *Bulletin of the National Society of Acclimatization of France* on January 1, 1940.

Magnetization may seem like a simple practice, but it involves many technical details. For example, a magnet is not just a magnet—it has power, polarity, and can be made of ferrite or neodymium, natural or artificial. There are basic parameters that are essential to understand in order to master the fundamentals of these unconventional methods and their applications.

The range of techniques using magnetism can then be deployed to enhance plant and crop growth, seed germination, and the resistance of vegetation.

For example, study No. 67 from the book *The Magnetic Pulse of Life 2*, Geomagnetic Effects on Terrestrial Life by Alan Cruise shows that a treatment of 1.5 Gauss applied to grape clusters resulted in *visible and significant* effects on the roots. Therefore, there are indeed practical and useful applications in agriculture, market gardening, orchards, and vegetable gardens.

The following section aims to guide you toward autonomy in your own research and experiments. There is not just one way to apply these practices, but rather an almost infinite number of methods. Let's get started!

The Earth's Magnetic Field

The attractive force of magnets has been known since ancient times, but it wasn't until around the **year 1000/1100** that the Chinese developed «compass» systems for navigation. The understanding of magnetism began to emerge in Europe as early as the **12th century** with the introduction of the magnetic compass. The compass itself was only created on this continent around **1300** by Flavio Gioia.

The study of modern magnetism was further developed by William Gilbert (**1544–1603**). In **1600**, he published On Magnetism, the Magnetic Body, and the Great Magnet, Earth, which is considered one of the first works dedicated to magnetism. W. Gilbert was a pioneering contemporary scientist who thoroughly developed a theory describing the Earth's global magnetic characteristics. He analyzed the planet as a colossal magnet, explaining the rules of attraction and repulsion, as well as the magnetization of iron within a magnetic field. He demonstrated how a compass in a magnetized area always points to the same location.

The famous mathematician and physicist René Descartes (**1596-1650**) also left his mark on the history of magnetism. He provided physical and mechanical explanations for many related phenomena. Later, Carl Friedrich Gauss was the first to determine the magnetic value of the Earth. Using the unit of measurement that now bears his name, *Gauss*, he conducted detailed statistical studies, notably through the establishment of a complex network of magnetic observatories distributed around the globe.

Today, archaeomagnetism studies the imprint of the Earth's magnetic field in ancient objects. For example, pottery from the 2nd century, during its firing, would have retained clues about the intensity of the Earth's magnetic field at that time. This can be observed by analyzing the arrangement of iron oxides contained in the fired clay.

It seems that the Earth's magnetism is trying to communicate with us...

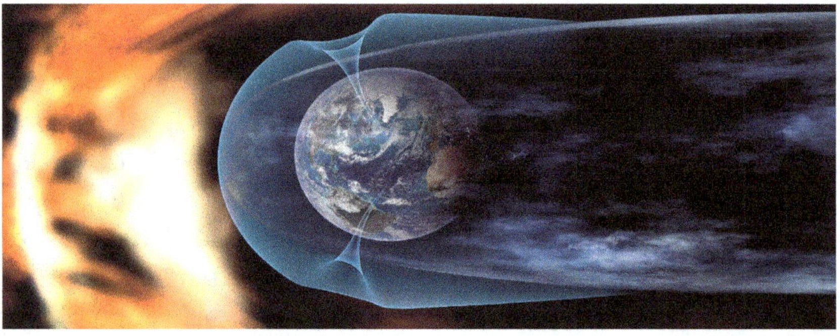

Figure 55 : Deformed Earth's Magnetic Shield Under the Action of the Sun.

The Earth's magnetic field extends well above the ionosphere, the upper layer of the atmosphere.

This protective shield originates from intense activities within the planet. At the Earth's center, there are two cores: the solid inner core and the liquid outer core, primarily composed of molten iron and nickel. A self-regulating dynamo mechanism between these cores, combined with convection movements in the liquid part, creates the magnetic dome that protects all life on Earth.

Figure 56 : The Earth and Its Dynamo-Magnetic Core.

The Earth's magnetic field can be compared to that of a bar magnet, with two polarities (dipoles), magnetic north and south. However, the axis of this enormous magnet is not exactly aligned with the Earth's geometric center, but rather offset from it (see figure 57). Similarly, the magnetic poles are not precisely located at the geographic poles; they are displaced and can move up to several tens of kilometers per year!

Additionally, the geographic poles are different from the magnetic poles—they are inverted! This inversion is due to an international convention: geographic north is actually magnetic south, and geographic south is magnetic north.

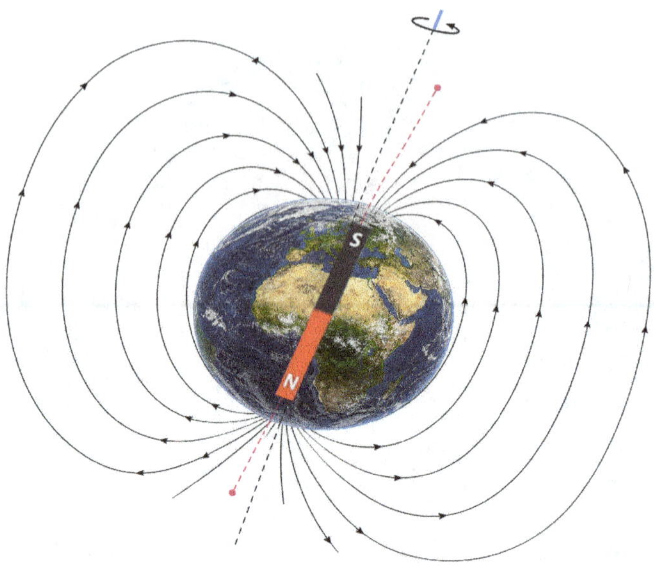

Figure 57 : Magnetic lines extend from geographic south to geographic north, but the magnetic poles are inverted relative to the geographic poles.

Furthermore, it has been established that the intensity of the Earth's magnetic field has been decreasing for the past 200 years. The impacts on life could be numerous. Indeed, changes in global magnetism have had, and will continue to have, effects on life on Earth. This is not a cause for alarm but rather an observation. In any case, electroculture can act locally on the magnetic aspect in fields, orchards, gardens, and vineyards. If global issues arise due to these decreases, solutions can always be implemented by intensifying local micro-magnetic fields that are essential for proper plant function.

The effects on life are already clearly visible. Over the past decade, I have translated numerous articles, books, and scientific studies to better understand the electromagnetic phenomena related to the development of living organisms. Science is a wonderful tool—it allows for concrete and rational approaches in research.

In the following two tables, I have condensed translations of studies from the book *The Magnetic Pulse of Life 2, Geomagnetic Effect on Terrestrial Life* by Alan Cruise (**2016**). This substantial 500-page book compiles a wealth of studies. For example, the 15 studies in the two tables are extracted from the 103 studies in Chapter 14 on plants (and there are many more!).

For each study, you will find the magnetic intensity of the treatment in mT (millitesla) and Gauss, the studied plant(s), the duration of the treatment, and the observations. Analyzing these allows us to draw conclusions:

1/ There are indeed observable effects (or none) on the plants.

2/ Treatment durations can vary and influence the results.

3/ Treatment intensities differ,

This helps us better understand and refine magnetic treatment techniques for plants, improving positive results. The goal is to adapt the choice of magnetization, treatment duration, and intensity based on the crop and desired outcomes. There is still so much to discover...

STUDY No.	MAGNETIC FIELD INTENSITY (mT / Gauss)	PLANTS
1	1 mT / 10 Gauss	*Koelreuteria paniculata* (Golden rain tree)
4	400 mT / 4000 Gauss	*Triticum* (wheat, barley and wild oats)
14	125 mT / 1250 Gauss	*Lycopersicon esculentum* (tomato)
15	500 mT / 5000 Gauss	*Zea mays* (corn)
17	0.05 to 500 mT / 0.5 to 5000 Gauss	*Brassica juncea* and *Amaranthus caudatus*
29	50 mT / 500 Gauss	*Triticum aestivum* (common wheat)
41	105 mT to 250 mT / 1050 to 2500 Gauss	*Oryza sativa* (rice)
45	0 to 10 mT / 0 to 100 Gauss	*Lactuca sativa* (lettuce)
48	0 to 10 mT / 0 to 100 Gauss	*Zea mays* (corn)
56	88 to 96 mT / 880 to 960 Gauss	*Fragaria × ananassa* (strawberry)

STUDY No.	EXPOSURE DURATION	Observations
1		increased germination.
4		viable effects on seeds soaked in water before treatment.
14	15 min	increased growth and development. Maximum effect starting after 15 minutes of seed exposure.
15		treatment of radicles with up to +2.7 % growth.
17		growth and development +25% for both plants at high intensities and +56% at low intensities for belladonna.
29		best results for seeds germinated 10 days after exposure. 30% increase in harvest yield.
41		seeds watered with magnetically treated water = improved germination rate.
45		increased water absorption. Affects osmotic pressure.
48		stimulates root development, germination, and increases root length and weight.
56		more leaves, roots, and fruits (maximum effects at 96 mT).

STUDY No.	MAGNETIC TREATMENT INTENSITIES in mT AND Gauss	PLANTS	EXPOSURE DURATION	OBSERVATIONS
60	100 mT / 1000 Gauss	*Vicia faba* (Fava bean)	Seed treatment	Seed treatment increases germination rate and precocity by 2 to 3 days.
63	80 to 120 mT / 800 to 1200 Gauss	*Lycopersicon esculentum* (Tomato)	Seed treatment	Seed treatment increases the length and weight of roots and stems, and fruit yield.
67	0.15 mT / 1.5 Gauss	*Vitis vinifera* (Grapevine)	5, 10, 15, and 20 min	Treatment of clusters = visible and significant effects on roots.
87	250 mT / 2500 Gauss	*Helianthus annuus* (Sunflower)	1 h	Improved integrity of the seed coat, germination, and seeding length.
88	80 mT / 800 Gauss	*Solanum tuberosum* (Potato)	Faster germination	Faster germination rate, increased plant weight.

Figure 58 : Scientific Articles on the Effects of Magnetic Treatments on Plants and Seeds

Figure 58 : Scientific Articles on the Effects of Magnetic Treatments on Plants and Seeds

For comparison, the intensity of the Earth's natural magnetic field at the surface ranges between 33,000 and 70,000 nT (nanoTesla) (see the article from *Futura Science*). When converted from nanoTesla to Gauss, this translates to 0.33 to 0.7 Gauss.

1 Tesla = 10 000 Gauss

1mT = 10 Gauss

100 000 nT = 1 Gauss

It is important to understand the Tesla/Gauss conversion. Magnetic intensity measurements are often expressed in either *Tesla* or *Gauss*.

The treatments applied in the 15 previous studies can therefore be up to, or even more than, 1,000 times stronger than the intensity of the Earth's natural magnetic field.

DESCRIPTION AND USE OF MAGNET(S)

What is a Magnet?

A magnet is a mechanically manufactured object made from compressed ferrite, to which magnetic properties have been imparted. During its production, a very powerful magnetic field is applied for a specific purpose, resulting in *axial magnetization* that creates a magnetic north pole and a magnetic south pole. This process gives us a magnet! It will then emit magnetic lines that form a local magnetic field, similar to the Earth's, but on a much smaller scale.

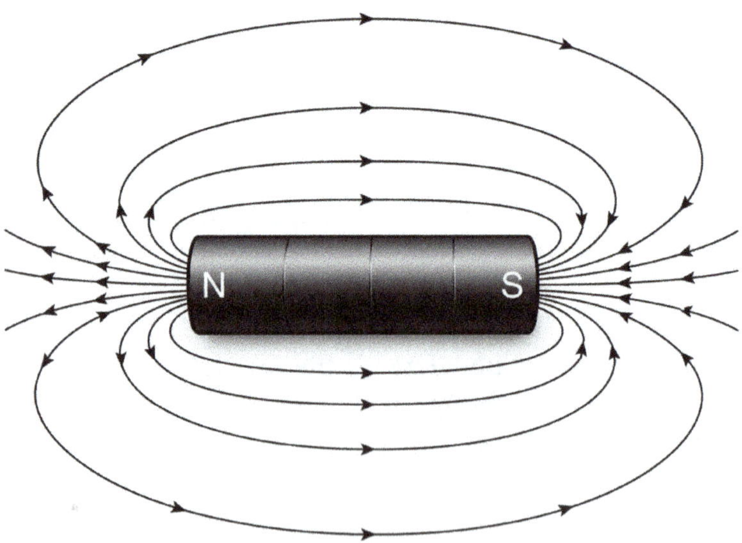

Figure 59 : Schematic of a ferrite magnet with magnetic lines extending from one pole to the other. The magnetic action is localized.

One of the Disadvantages of a Ferrite Magnet: The action of a ferrite magnet is localized; typically, the magnetization of 1,000 Gauss is concentrated close to the magnet. As you move just a few centimeters away, the strength drops significantly... and continues to decrease until it becomes negligible.

But how can we increase the magnetization area of a magnet?

Thanks to the talented Justin Étienne Christofleau, a French engineer and inventor of the **20th century**. He was a member of the Association of Inventors of France and was also awarded the title of Knight of the Order of Agricultural Merit.

A newspaper article in *Siroco: the true journal of the Youth of France* from **October 23, 1943**, discusses one of his inventions. Here's an excerpt (**see images and articles on the following pages**).

Figure 60 : Telluric Fertilizer Created by Christofleau

«- You firmly attach this... this device, look! I have one here, waiting in the corner of my desk. You see, it's about 24 inches long, with two points, a sort of half steel tube adorned with five blades. You place it at the head of the field, oriented South-North—this is essential. Then, you run parallel iron wires 10 feet apart, which can be up to 0.6 miles long. These wires must also be oriented South-North and buried deeper than you can plow or harrow. Because once the installation is in place, you won't have to worry about it for 50 or even 100 years—there's no limit! The work takes care of itself.

What work?

The electrification of the soil in the field! My steel device attracts the magnetic flow continuously coming from the South Pole, condenses it, and from there, electricity radiates along the iron wires to the northern exit of the field, activating everything in the soil that nourishes the wheat. Wheat or whatever else you want: barley, corn, potatoes, lettuce, peas. It's a true fertilizer, I tell you! But it costs nothing and is far more powerful than many others!»

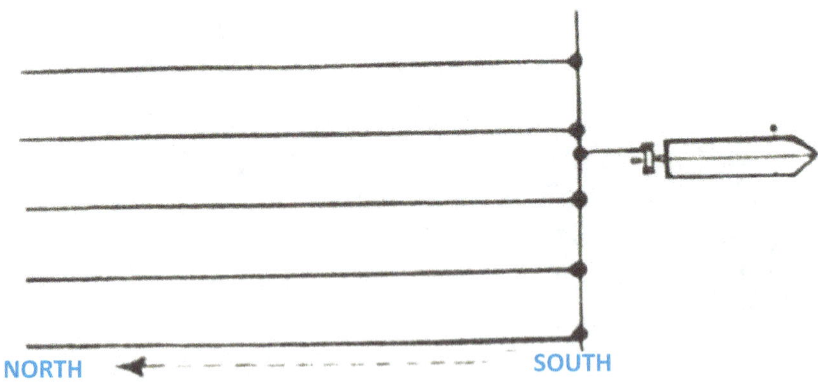

Figure 61 : Fertilizer connected to iron wires to transmit magnetism.

There are several important points to note from this excerpt :

1/A piece of cast metal can :

> *Capture the earth's magnetism.*

> *Transmit magnetism to iron wires.*

> *Act as a magnet!*

2/ And :

> *The installation must be aligned with the Earth's magnetic field!*

One day will come when
BY THE SMOKE
extracted from the mi

Of nothing, and which is so simple and yet so powerful, was it a Frenchman who invented it — CHRISTOFLAIT. I wonder what you think of it, you who, in Seine-et-Oise, have meadows where the grass is two metres high, and yet only grows four metres, and perfectly dies, whereas it could not die in that valley; peasants there measure twenty centimetres around the waist, and boys of three metres and fifty centimetres in circumference !

Fame ! All this because of these little steel machines that he installed in his garden.

He must believe it ! For a gram of smoke is more valuable than it has been deposited on that land for a quarter of a century. Naturally, you might say that Ile-de-France is just a fragment found in Paradise on earth; but it is only a small part of the world that will try fertilizatiom of plants by the magnetism of the earth:

ur fields will be fertilized
IN IRON WIRE
nes of the South Pole

Tonkin, in Italy, in Switzerland, in Australia, in Venezuela, in Spain. What do you think, for example, of this experience done around Marseille on a vineyard plan for old electric traction? No at all! A treatment of electroculture, and the following harvest was covered with superb grape bunches!

— You put wine in my mouth! But then, a question, old man, why the smoke-in-iron-wire system — does it respond so little?

— First, my dear old man, because it is still little known. And then, it's very simple for our people today who don't believe that engines driven by infections with manometers, levers, brakes, odometers, gearboxes, and everything in between! You think: here is a simple steel network buried in a steel bar buried in the soil — and the South Pole does the rest!

"If only you could know," you dream, that if the whole world's fruits were heavier, the wheat more abundant, life much happier and peoples might be less at war..."

J.-L.-D.

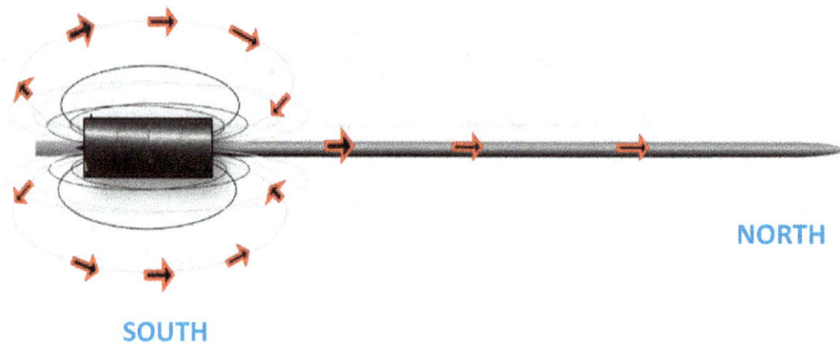

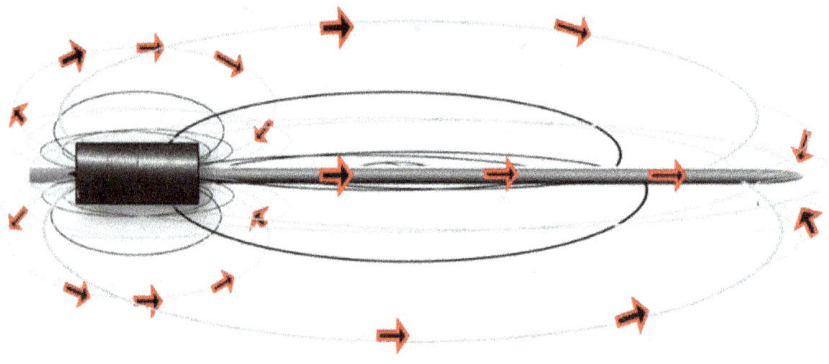

Figure 62 : Connecting a magnet to a ferromagnetic wire will cause the wire to become magnetic itself.

Connecting the magnet to a ferromagnetic iron bar or wire will extend its magnetism.

> **Important** : The magnet must be placed on the geographic south side, and the wire should run along the geographic north side.

The application of this technique can be used to fertilize rows of vines, alignments of kiwis from north to south, energize seeds or water, enhance raised garden beds, or even support the trellis for your favorite climbing plant... The possibilities are endless!

This creates a locally amplified natural magnetic field, connected to the Earth's magnetism. The cultivated plants generally become more vigorous and resistant to potential diseases. Installing magnets can help promote the natural balance of the soil and enhance fertility around a galvanized wire in a natural way.

However, I emphasize once again that there is no «*miracle recipe.*» These are paths of reflection to explore and experiment with, always aiming for positive outcomes. Each context is different and requires personal analysis. For example, a certain magnetic strength (in Gauss) might affect one plant differently than another. There are so many parameters to consider...

What I recommend is starting small—experiment on a small plot before expanding to your entire space. This is one of the principles of permaculture: «*Apply on a small scale and with patience.*» Conduct your own experiments. Consider advice, but implement it through successive trials to find the best formula for your unique situation.

HOW TO IMPLEMENT?

Firstly, the primary goal is not necessarily to invest time, energy, carbon footprint, or money in installing iron wires in your soil or at home. The logical first step would be to use this technique if you already have wires present on your site. For example, if your vineyard is already well-oriented and has ferromagnetic wires, what could be simpler than adding magnets?

To begin your first experiments, you can place the magnets only on the geographic south side of the wire. If desired, you can also try adding more magnets along the wire to further enhance the magnetism. The possibilities for experimentation are limitless. Below, I propose a few examples, though not exhaustive, to inspire your applications.

Garden, Open Field, Vineyard, and Market Garden Greenhouse

In a vegetable garden with a permanent raised bed or a permaculture-style plot, it is possible, during the design phase, to integrate a galvanized wire with one or more magnets along the wire. The magnetic antenna should then be placed on the geographic south side. The system must always be oriented along the axis of the Earth's magnetic field. It is preferable to bury everything 8 to 12 inches deep at the root zone level. You can also bury it deeper, up to 20 inches, or even lay the wire on the surface.

In the latter case, it will be easier to move if you need to use a garden fork or work the soil for planting. Indeed, if the wire's north/south magnetic alignment is displaced when buried, it will no longer operate optimally or may not work at all. The alignment with the Earth's magnetic field is crucial to stimulate the ferromagnetism of the wire in connection with the magnet.

Figure 63 : Trench for a Series of Magnets and Ferromagnetic Wire

Figure 64 : Ferrite Magnets on Cultivation Beds

Optimizations in the garden

It is clear that the soil must be alive and protected by mulch to better stimulate and preserve life. Covering the soil helps retain moisture and enhances the magnet's conductivity for electrical and magnetic energies. Paramagnetic basalt can also be spread and incorporated into the soil to amplify the energetic effects (**see the book *Energy Towers and Paramagnetic Basalt* by Loïc Etcheberry**).

In a greenhouse: If the orientation is ideal, as you now understand, simply add a series of magnets to the galvanized wires inside. Some market gardeners practice companion planting, and sometimes climbing plants grow on the walls; this technique can assist in those cases.

Figure 65 : Ferrite magnets on a market gardening greenhouse.

For the vineyard : If the support structure is properly oriented north-south, simply place the magnets on the ferromagnetic cables. You can experiment by placing a series of magnets on one or more wires along the row, and even at several points along the wire—there are no limits to research! In my opinion, it would be best to conduct comparative tests on one row during the first year before expanding the technique to a larger area. You can also experiment with different magnetic intensities in Gauss.

Figure 66 : Ferrite magnets on a row of vines.

Seed Dynamization

Justin Étienne Christofleau also developed another intriguing technique in his time. It involved a special box called the *electro-magnetic-celestial conserver*, designed for «electrifying seeds and fruits.» This passive system was used for food preservation and seed dynamization. The box needed to be aligned with the Earth's magnetic field to function passively. Inside, it contained soft iron bars, zinc, copper, and magnets! With the correct orientation, all the components would magnetize to interact with the contents inside.

These facts can be verified in Patent No. 570,286 filed by Léonie Reboisson, secretary to Justin Étienne Christofleau, and granted on **January 15, 1924**.

What this patent states: «*A box whose interior walls and lid are equipped along their entire lengths with soft iron bars, or any other metal suitable for capturing and conducting terrestrial magnetism and telluric currents, provided they are exposed in the direction of the compass needle.*» «*The aim is to form a magnetic field between its bars to preserve the fruits placed in the magnetic field created by the box, and to give the seeds within it the possibility of charging with natural electricity, thus acquiring greater germinative power and vitality.*»

I will elaborate further on these *dynamization* techniques for seeds in another book, as there is much more to say on the subject.

IMPORTANT : What orientation for the magnet ?

As explained earlier, a magnet has two magnetic poles (north and south).

It is therefore crucial to position the magnet in the correct orientation! The magnetic lines should not oppose the Earth's magnetic field. To verify this, bring a compass close to the magnet; the needle should not reverse. If it does, the magnet needs to be adjusted. Once correctly aligned, the magnet will be connected to the Earth's magnetism.

However, if the compass needle reverses when brought near the magnet, simply adjust its orientation. Once aligned, secure it in place to prevent movement, or make adjustments as needed when conducting experiments.

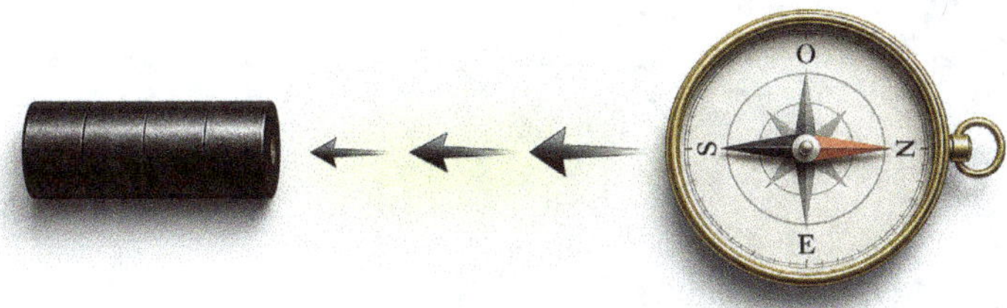

Figure 67 : To avoid mistakes, place the magnet in the same orientation as the compass.

WARNING! When bringing the compass close to the magnet, it should be done slowly and along its axis of magnetization.

Mickaël from Limoges testifies.

Mickaël practices permaculture and has been experimenting with electroculture for several years. Passionate about gardening and nature, he ensures that his soil is always covered in his vegetable garden. He also sows green manures between crops and has applied paramagnetic basalt on his plots (2 handfuls per square meter). Over time, he has installed several aerial and magnetic antennas at home, including one that reaches 39.4 feet in height! His greenhouse is equipped with a series of magnets connected to galvanized wire, all oriented along the Earth's magnetic field.

The results are impressive: a sweet potato plant producing 19.4 lbs, a beet weighing 15.7 lbs, and a sunflower head measuring over 23.6 inches in diameter, with photos to back it up.

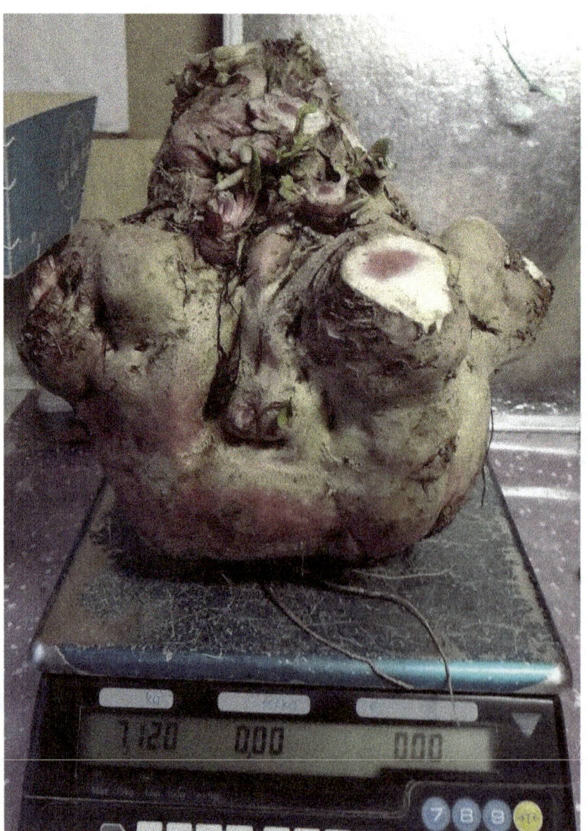

Figures 68 et 69 : A beet weighing 15.7 lb and a sunflower head measuring over 23.6 inches in diameter.

Bio-electronics Vincent

Louis Claude Vincent
Interests of BEV
How to measure?
Applications

This section is intended for those who wish to become familiar with precise measurement tools to assess the electromagnetic influence of electroculture systems on their cultivation environment. Although these concepts may seem technical or complex, they provide a solid foundation for moving beyond subjectivity and providing tangible evidence. Here, we will cover *Bioelectronic Vincent* (BEV), a method developed by *Louis Claude Vincent*.

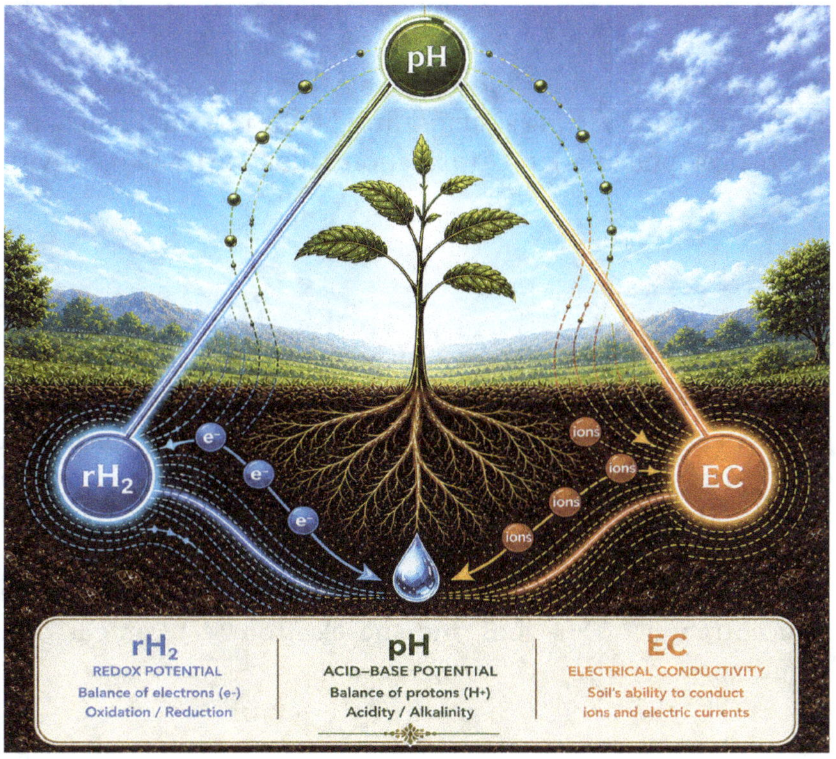

My goal is to make these concepts more accessible, as they represent one of the essential keys to understanding life and fertility. If this introduction interests you, I encourage you to explore this fascinating field further.

WHO IS LOUIS CLAUDE VINCENT?

Louis Claude Vincent was an engineer who graduated from the École supérieure des Travaux publics, born on **January 10, 1906,** in Puy-en-Velay, France. He co-founded the Centre for Bioelectronic Research with pharmacist Dr. Jeanne Rousseau and taught at the School of Anthropology in Paris. Together with German physician Dr. Franz Morell, he also established the International Society of Bioelectronics. Vincent developed a clear and systematic method known as «Bioelectronics,» which allows for the analysis of living systems through the measurement of pH (Hydrogen potential), rH2 (oxidation/reduction), and resistivity.

Interests of BEV (Vincent Bioelectronics) ?

These precise measurement techniques offer a complementary understanding of living systems, soil, diseases, plants, and crops. This approach focuses on their atomic (electromagnetic) composition and interactions with the environment. It is based on the study of negative electron charges (rH2) and positive magnetic charges of protons (pH), defined by measuring the concentration of H+ ions and the electron (e-) content of a solution.

Based on the results obtained, a classification into **four main environments** is established, according to Éric Petiot's method:

1/ Rich in protons and low in electrons = «Acid-Oxidized,» conservation force.

2/ Low in protons and electrons = «Alkaline-Oxidized,» degradation force.

3/ Rich in protons and electrons = «Acid-Reduced,» vital force.

4/ Low in protons and rich in electrons = «Alkaline-Reduced,» putrefaction force.

According to Éric Petiot, these four electromagnetic grounds correspond to certain natural laws. They represent the forces of conservation, degradation, vitality, and putrefaction.

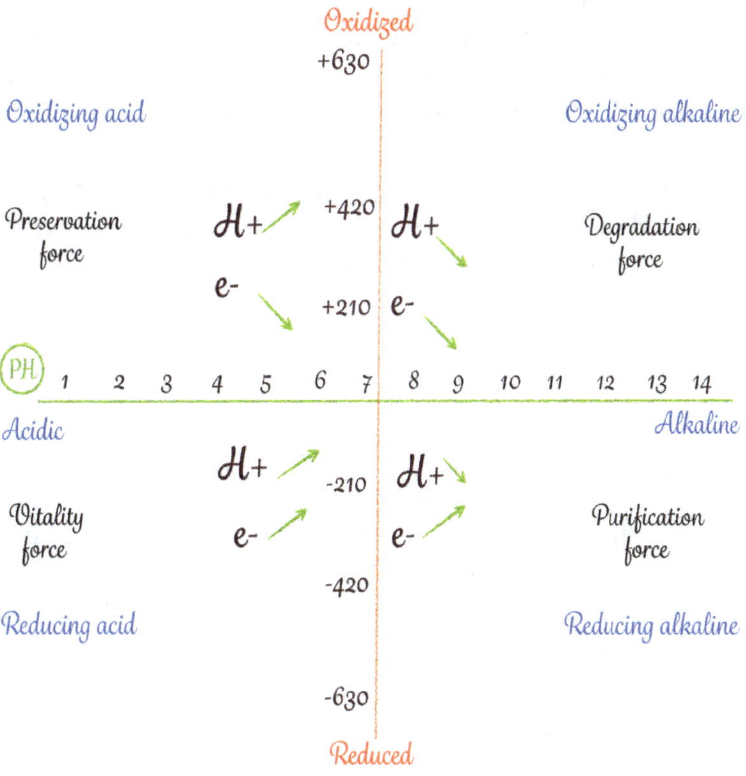

Figure 70 : French table of bioelectronic classification according to Éric Petiot in his book *L'agriculture énergétique*.

How to measure? Three essential parameters:

pH (Hydrogen Potential): Measures the richness or scarcity of protons (H+ ions). The more acidic a liquid is, the richer it is in protons; the more basic it is, the poorer it is in protons. Acidity (below 7), basicity (above 7), or neutrality (7) is defined by a value index ranging from 0 to 14.

Measuring Device: pH meter.

Redox (rH2): Measured in millivolts (mV), it indicates the richness or scarcity of electrons (e-). Simply put, the higher the measured value, the fewer electrons are present; the lower the value, the more electrons there are. A poor medium is termed «oxidized,» while a rich medium is called «reduced.»

For example, two seemingly similar tomatoes may differ significantly in electron richness, which impacts their nutritional quality and health benefits. Electroculture can help enrich vegetables and fruits with electrons.

Measuring Device: Redox meter.

Resistivity: Measures a solution's ability to resist the passage of electric current (the inverse of conductivity) by assessing the concentration of electrolytes (molecules, ions, minerals). For instance, a very low concentration of minerals will impede the current's flow.

Measuring Device: Conductivity Meter.

Figure 71 :
Loïc with a bioelectronic measuring device of the Consort type.

CONCRETE APPLICATIONS

If a tree is affected by a disease (the sick tree) that thrives in an Alkaline-Oxidized environment, a rebalancing treatment from the Acid-Reduced category can be applied. For instance, this could involve using a natural active ingredient that aligns with this opposite environment to counter the disease. However, it is crucial to ensure that the plant has already activated its defense mechanisms before applying such treatment to avoid unintentionally benefiting the disease. Indeed, plants defend themselves through superoxidation against pests, so no reducing products should be directly applied during an active pest attack.

> *It is possible to begin by conducting an energetic diagnosis to assess the plant's condition and then apply targeted, suitable measures to restore its balance.*

It is widely understood that diseases, pests, or health issues often arise in an unbalanced environment or ecosystem. Visible symptoms typically stem from a deeper underlying condition that may not be immediately obvious.

For instance, if scale insects infest your houseplant, the root cause likely lies in an imbalance, such as insufficient nutrients in the soil or inadequate light in the room. In this case, the scale insects are not the true cause of the problem but rather a symptom of a deeper issue.

Moreover, the presence of a pest often serves to eliminate what is weak, contributing to natural selection and the overall well-being of the ecosystem. Adopting a holistic perspective remains crucial when assessing the health of an organism or environment. This approach aims to restore energetic balance, which frequently leads to the resolution of diseases.

> *The plant constantly strives to achieve both internal and external balance of its bioelectronic parameters in order to maintain its health.*

When the Eh (electron concentration, «e-») and pH (proton concentration, «H+») are at optimal levels, the plant's functioning within its environment is efficient. This means that the stabilization of certain physiological characteristics in plants is achieved through an internal harmony of Eh-pH.

Two examples from Olivier Husson:

Plants **«modify these parameters in the rhizosphere to ensure homeostasis in their cells,»** which means that plants can adjust the concentration of electrons (e-) and protons (H+) around their roots, helping them maintain the overall balance of their organism.

«The optimal pH for most cultivated plants ranges between 6.5 and 7, with favorable growth conditions from 5.5 to 7.5. The ideal redox potential (Eh) for soil is around 400 to 450 mV; when it drops below 350 mV, plant growth rapidly declines. The upper limit for Eh is difficult to pinpoint, but at a pH of 6.5 to 7, a potential between 450 mV and 500 mV becomes unfavorable, potentially leading to mineral deficiencies, heavy metal toxicity, and increased pathogen development.»

Indeed, this reasoning is technical, but it clearly demonstrates that serious and concrete research has been conducted in agriculture on this subject. Additionally, many factors can disrupt or modify these bioelectric potentials in the soil, such as seasonality, tillage, moisture levels, irrigation water quality, time of day, and even electroculture systems!

By leveraging these insights and progressing toward the use of such measurement tools, it becomes possible to adjust the parameters more effectively. This allows for a more detailed understanding of the actions of specific techniques and their impact on the environment (or lack thereof). The ultimate goal, for instance, is to enhance agricultural or horticultural production.

The classification table by Olivier Husson

This table differs slightly from the one discussed earlier, particularly because it includes, diagonally, an axis labeled «pe + pH.» I won't delve into the technical details here. However, what is particularly interesting, and easy to understand for simplification purposes, is that in this example, insects, viruses, fungi, nematodes, healthy plants, and other factors are positioned at specific points in the table based on pH, Eh (mV), and pe + pH values.

Eh-pH and Plant Health

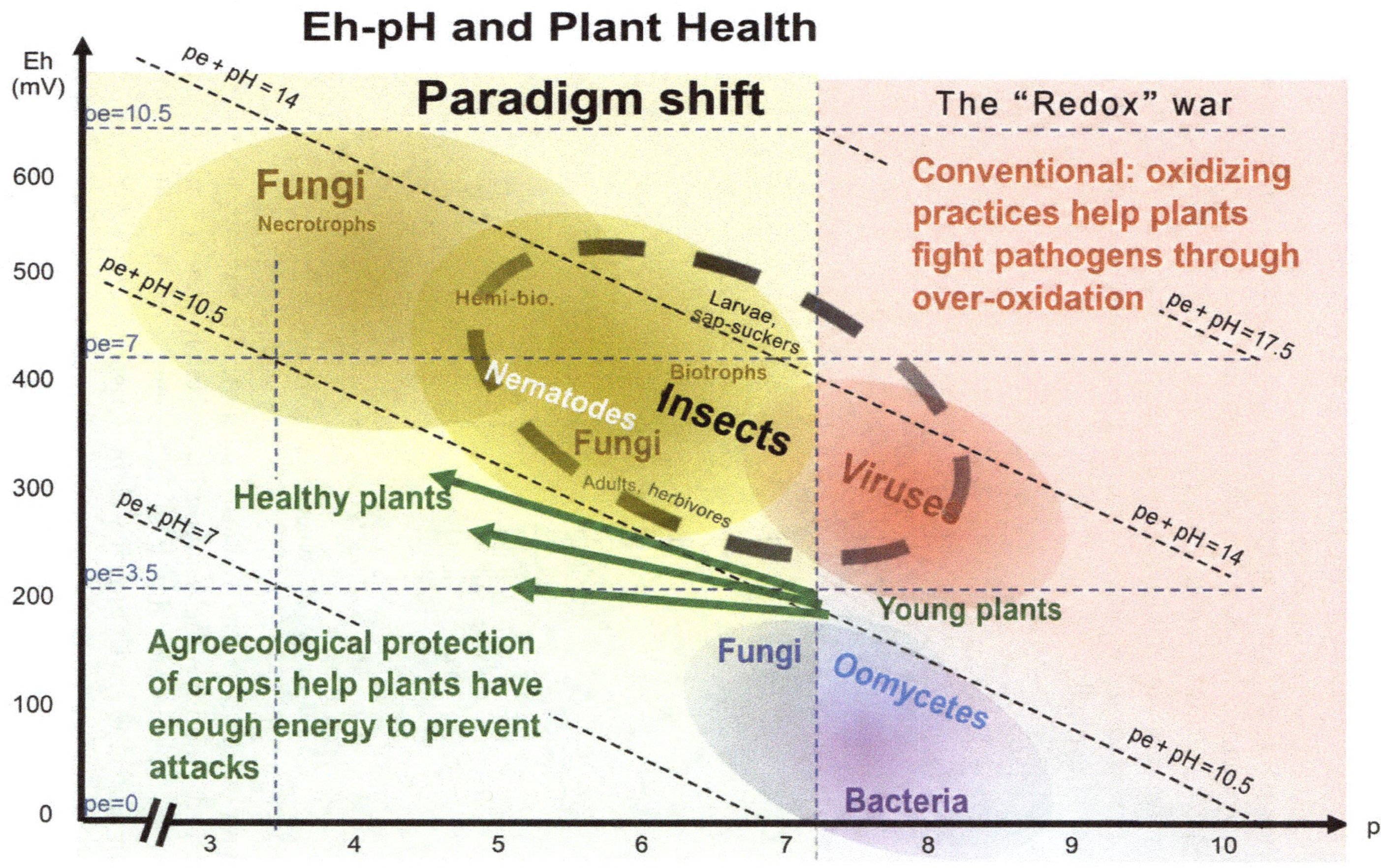

Figure 72 : French bioelectronic Classification Table Used by Olivier Husson

Healthy Plants Evolve in a Well-Defined Environment

Remember, at the beginning of this book, I mentioned Beckensteiner. In **1882**, he was cited in the Lyon Agricultural and Viticultural Gazette under the title «*Striking Down of the Phylloxera.*» He employed an ingenious electroculture system known as the geomagnetifier to effectively combat a sap-sucking insect: the phylloxera.

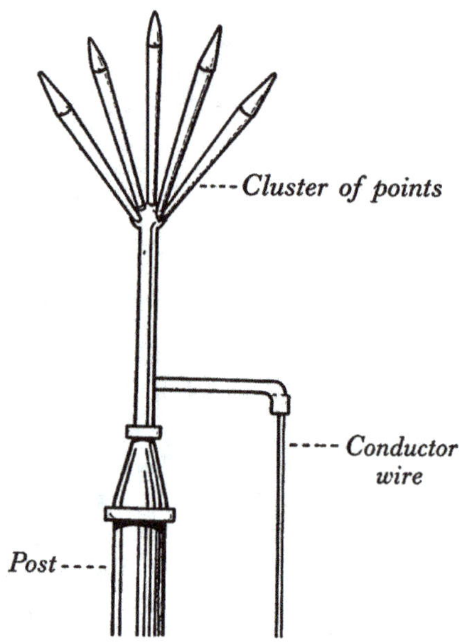

Beckensleitner's Geomagnetosphere.

Figure 73 : Geomagnetifier of Beckensteiner.

In my opinion, it's reasonable to assume that the «geomagnetifier» helped modify the surrounding environment, likely by increasing the availability Aef electrons near the plants. With this shift, the environment was no longer favorable for the pest but instead restored to one that supports healthy plant growth.

Everything is connected...

Conclusion

Electroculture, or Energy Agriculture, is a fascinating field that presents both challenges and opportunities for the 21st century. With the significant discoveries in electromagnetism, the potential to improve agricultural and gardening practices is now within our grasp.

Aerial and magnetic antennas, along with Vincent's bioelectronics and water dynamization, are powerful tools that can significantly enhance plant growth, improve food quality, and reduce the reliance on chemical inputs that our modern agricultural systems have become so dependent on.

Delving into the history and techniques of electroculture reveals almost limitless possibilities for application, and as we've seen, the results can be truly impressive. Historical examples demonstrate almost unbelievable plant growth, increased resistance, and enhanced quality. Water dynamization, too, has shown unexpected benefits for crop vitality, and magnets have been found to assist in seed dynamization.

Yet, despite all these positive outcomes, it's essential to recognize that these methods are still evolving, with much more to be learned. There are no one-size-fits-all solutions. Instead, there are promising pathways that need to be adapted to individual contexts. This is why ongoing research, experimentation, and the sharing of knowledge and results are vital to further advancing this vast and liberating field.

For my part, I will continue my mission of research and observation, always sharing my findings so that we can collectively move towards a more ecological, sustainable, and supportive future.

As farmers, scientists, winemakers, gardeners, market gardeners, fruit growers, permaculturists, and more, we each have a role to play in exploring and applying this ancient knowledge.

Electroculture is a constantly evolving field, with new discoveries emerging all the time. Let's continue to experiment and share our knowledge, pushing forward together to unlock the full potential of this fascinating discipline.

By pooling our expertise and spreading our passion, we can contribute to a world where vegetables are of superior quality, larger, and more natural, fruits are tastier and more nutritious, and water is cleansed of harmful chemicals.

Finally, I wish you all the best in your research and experiments, and I look forward to joining you on new adventures soon!

THANK YOU.

Loïc Etcheberry
www.permafutur.com

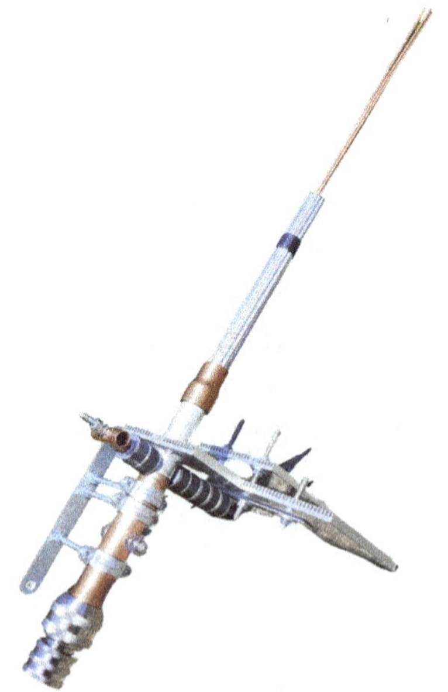

BIBLIOGRAPHY

BOOKS

ABBÉ BERTHOLON Pierre-Nicolas, «De l'électricité des végétaux, Partie III L'électrovégétomètre», 1783 / «L'électricité de l'atmosphère sur les plantes, des effets sur l'économie des végétaux, des «vertus médico et nutritivo-électrique» et des moyens d'applications utiles à l'agriculture», 1783.

BASTY Fernand, «Essais d'électroculture œuvres complètes», 1908, / «De la fertilisation électrique des plantes», 1911.

BARTHOLOMEW Alick, le «Le génie de Viktor Schauberger», 2003.

CASTELL Roger, «La bioélectronique Vincent, technique et méthode de santé naturelle pour augmenter votre vitalité», 2014.

CHRISTOFLEAU Justin, «Augmentation des récoltes et sauvetage des arbres malades par l'Électroculture», 1921 / «Électroculture», 1925.

CASTEL Roger, «La bioélectronique Vincent, Technique et méthode de santé naturelle pour augmenter votre vitalité», 2011.

CRUISE Alan, «The magnetic pulse of life 2, geomagnetic effect of terrestrial life», 2016.

DELMAS, CHAUZY, VERSTRATE & **FERRÉ**, «Atmosphère, océan et climat», 2007.

ETCHEBERRY Loïc, «Tours d'énergie et basalte paramagnétique», 2022.

LE MOUËL Jean-Louis & POIRIER Jean-Paul, «Une brève histoire du magnétisme», 2013.

LEMSTRÖM Karl Selim, «Electricity in agriculture and horticulture», 1904 / «L'aurore boréale» 1886.

MARTIGNAC Pierre, «Les vertus du magnétisme des aimants et de la terre», 1997.

PAULIN, «De l'influence de l'électricité sur les végétaux», 1892.

PETIOT Éric, «L'agriculture énergétique», 2014.

SILBERNAGEL A. et BASTY Fernand, «L'électroculture, technique agricole moderne», ré-édition 2023.

SILBERNAGEL A., «1er congrès international d'électroculture et des applications de l'électricité à l'agriculture», 1912 et 1914.

STEINER Rudolph, «Agriculture, fondements spirituels de la méthode bio-dynamique», 1999.

TAVERA Matéo, «La mission sacrée», 1969.

TRAMBLAY Benoit, «La cire d'abeille, une radio cosmique ?», 2014.

VINCENT Louis Claude, «Aperçu sur la bio-électronique», 1954.

VIOLET Marcel, «Le secret des patriarches», 1962.

ARTICLES

1882 «La gazette agricole et viticole de Lyon» title «Foudroiement du Phylloxera».

1890 «L'ami du peuple» May 11.

1891 «Le carillon valentinois» March.

1892 «Lyon Horticole» du January 15.

1894 «Le magasin pittoresque» January 1.
1894 N°620 «Génie civil, revue générale des industries française» April 28.

1911 «Bulletin de la société Royale de botanique de Belgique».

1912 «Congrès international d'électroculture» October 24 at 26, Reims.

1926 «Encyclopédie des nouvelles inventions des sciences et travail».

1930 «Rustica» August 17.

1932 «Paris-Soir», Alexis Danan, January 11.

1934 «Le magazine de Paris «La liberté»» November 3.

1937 Research Bulletin No. 210 of the College of Agriculture and Mechanic Arts of IOWA, Charles S. Dorchester «The effect of electric current on certain crop plants» March.

1940 «Bulletin de la Société nationale d'acclimatation de France» January 1.

1943 «Siroco, le vrai journal des Jeunes de France», N° 41 October 23.

1978 «Rustica» n°456 «Ils électrisent le potager…».

2000 «Fruits oubliés » n°2, « L'Électroculture, une méthode d'avenir ?».

2010 «Nexus» n°69, July-August 2010.

2016, Futura science, «Champ magnétique terrestre, qu'est-ce que c'est ?».

STUDIES / PUBLICATIONS

HUSSON Olivier, «Redox potential (Eh) and pH as Drivers of Soil/Plant/ Microorganism Systems, A Transdisciplinary Overview Pointing to Integrative Opportunities for Agronomy», 2013 / «Soil and Tillage research, Conservation Agriculture systems alter the electrical characteristics (Eh, pH and EC) of four soil types in France », 2017, « pH et redox en agriculture».

SIMPSON George «The electricity of rain and its origin in thunderstorms», 1910.

WHEATON Fredrick Warner, «Effects of various electrical fields on seed germination», 1968.

EMOTO Masaru «La mémoire de l'eau».

f Permafutur   ▶ Permafutur ⊙ Permafutur

Shop, workshops, electro-culture products and services :

🌐 **www.permafutur.com**

f Facebook group :
Électroculture avec Permafutur

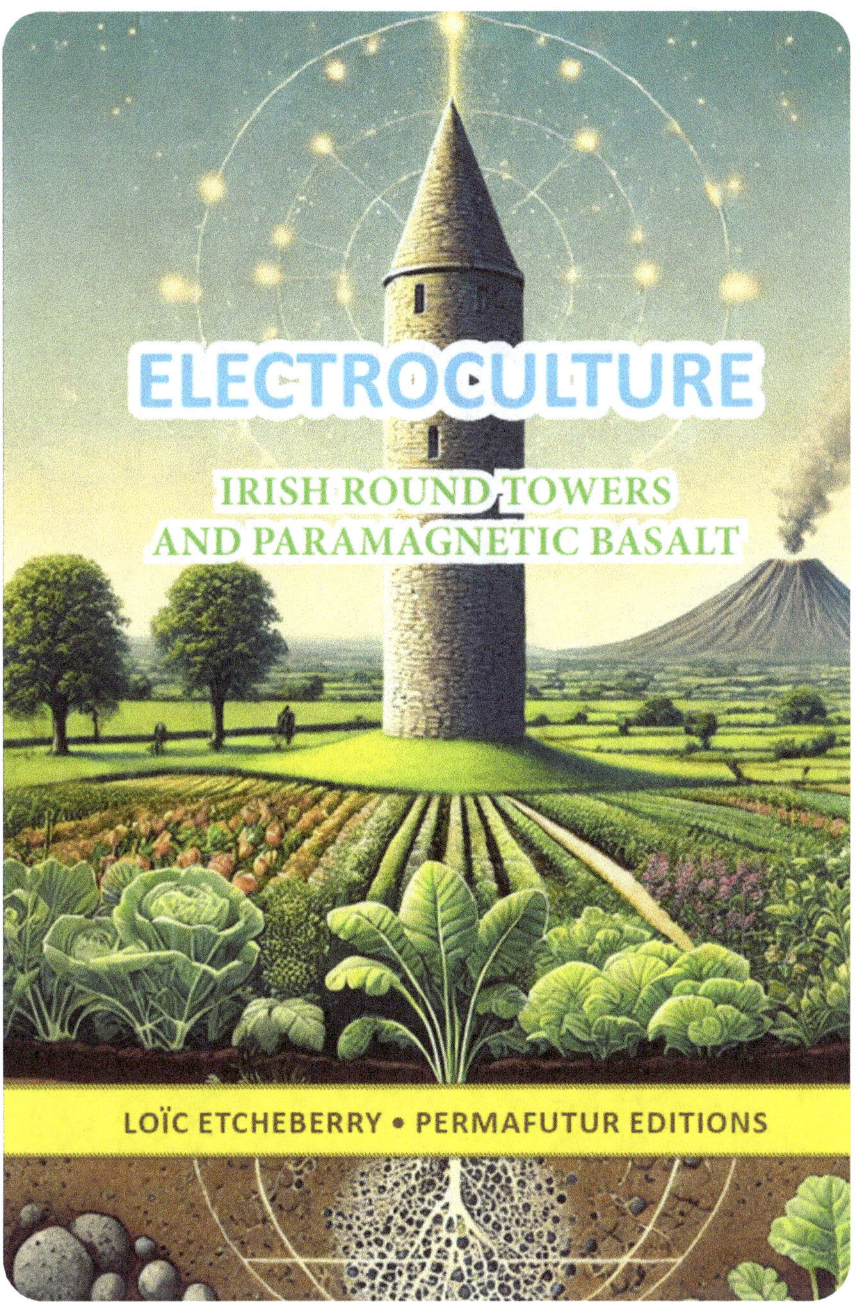

Other book
at Permafutur Editions